THE VISUAL
DICTIONARY *of the*
HUMAN
BODY

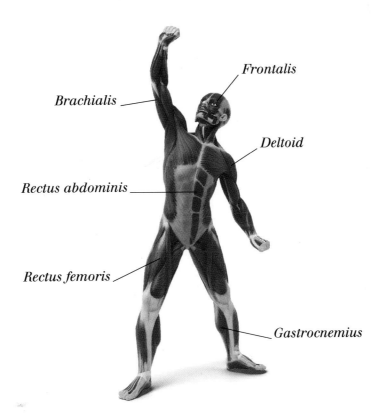

Brachialis

Frontalis

Deltoid

Rectus abdominis

Rectus femoris

Gastrocnemius

SUPERFICIAL SKELETAL MUSCLES

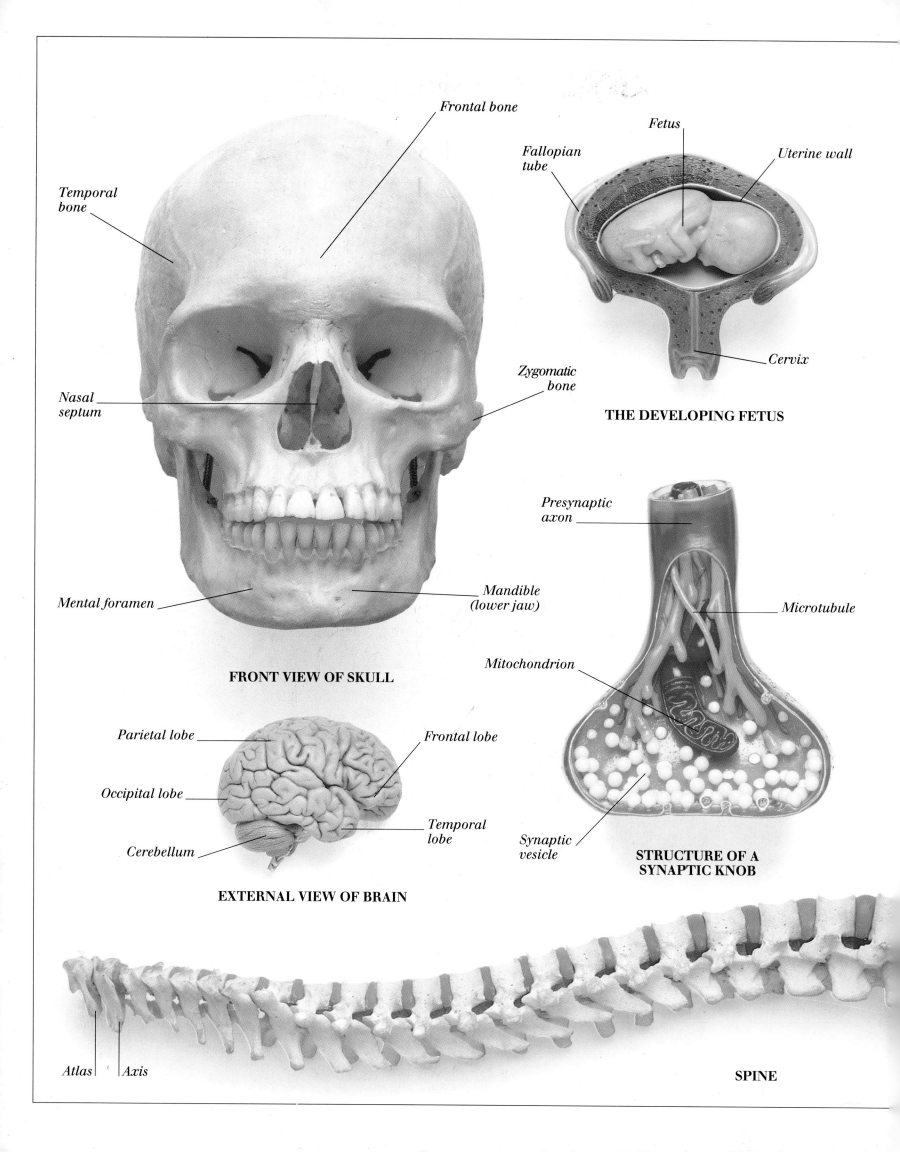

Frontal bone

Temporal bone

Nasal septum

Mental foramen

Zygomatic bone

Mandible (lower jaw)

FRONT VIEW OF SKULL

Fetus

Fallopian tube

Uterine wall

Cervix

THE DEVELOPING FETUS

Presynaptic axon

Microtubule

Mitochondrion

Synaptic vesicle

STRUCTURE OF A SYNAPTIC KNOB

Parietal lobe

Frontal lobe

Occipital lobe

Temporal lobe

Cerebellum

EXTERNAL VIEW OF BRAIN

Atlas

Axis

SPINE

EYEWITNESS VISUAL DICTIONARIES

THE VISUAL
DICTIONARY *of the*
HUMAN
BODY

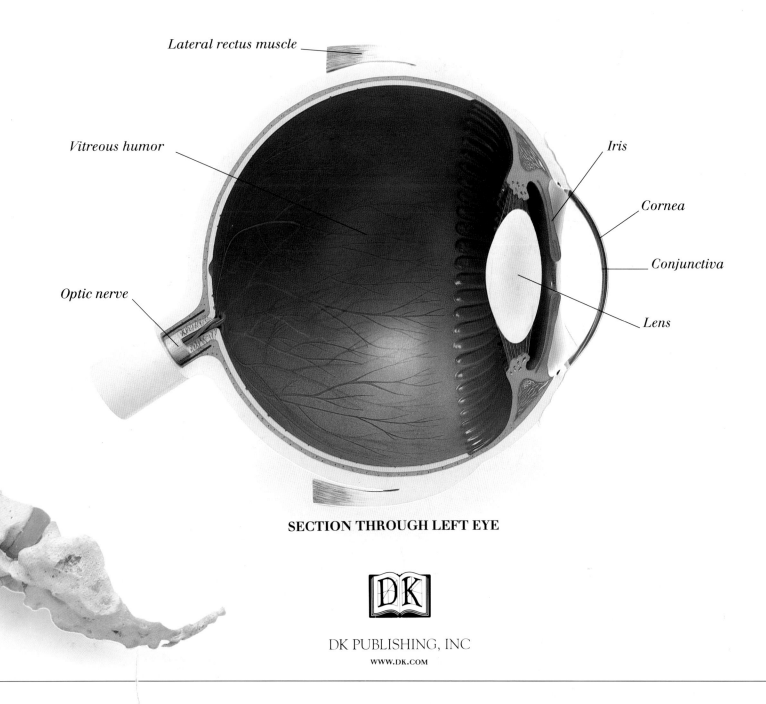

Lateral rectus muscle

Vitreous humor

Iris

Cornea

Conjunctiva

Optic nerve

Lens

SECTION THROUGH LEFT EYE

DK

DK PUBLISHING, INC

WWW.DK.COM

A DK PUBLISHING BOOK
WWW.DK.COM

PROJECT ART EDITOR BRYN WALLS
DESIGNERS DUNCAN BROWN, SIMONE END, NICKI LIDDIARD

PROJECT EDITOR MARY LINDSAY
CONSULTANT EDITORS RICHARD CUMMINS, FRCS, DR FIONA PAYNE, DR FRANCES WILLIAMS

SERIES ART EDITOR PAUL WILKINSON
ART DIRECTOR CHEZ PICTHALL
MANAGING EDITOR RUTH MIDGLEY

PHOTOGRAPHY PETER CHADWICK, GEOFF DANN, DAVE KING

PRODUCTION HILARY STEPHENS

SPECIAL THANKS TO THAD YABLONSKY

ANATOMICAL MODELS SUPPLIED BY SOMSO MODELLE, COBURG, GERMANY

Superior vena cava — *Aorta*

Right ventricle — *Left ventricle*

CIRCULATORY SYSTEM OF HEART AND LUNGS

FIRST AMERICAN EDITION, 1991

18 20 19

PUBLISHED IN THE UNITED STATES BY
DORLING KINDERSLEY, INC., 375 HUDSON ST.
NEW YORK, NEW YORK 10014

COPYRIGHT © 1991 DORLING KINDERSLEY LIMITED, LONDON

ISBN: 1-879431-18-1 (TRADE EDITION)
ISBN: 1-879431-33-5 (LIBRARY EDITION)

LIBRARY OF CONGRESS CARD CATALOG NUMBER: 91-060899

REPRODUCTION BY COLOURSCAN, SINGAPORE
PRINTED AND BOUND IN SPAIN BY ARTES GRAFICAS TOLEDO S.A.U.
D.L. TO: 61-2004

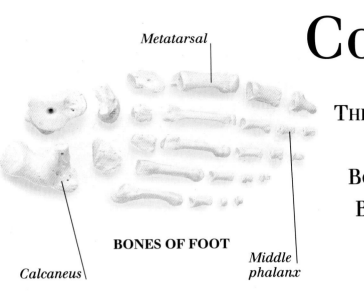

Metatarsal

Calcaneus

Middle phalanx

BONES OF FOOT

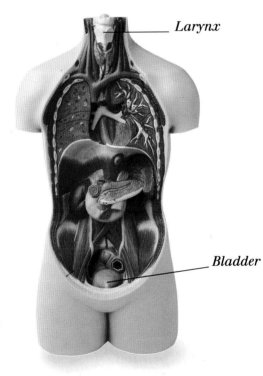

Larynx

Bladder

CHEST AND ABDOMINAL CAVITIES WITH SOME ORGANS REMOVED

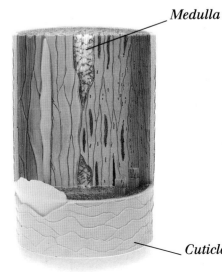

Medulla

Cuticle

SECTION OF HAIR

Contents

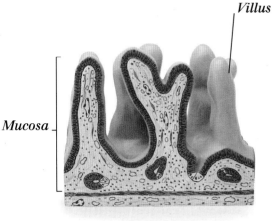

Villus

Mucosa

INTERNAL SURFACE OF JEJUNUM

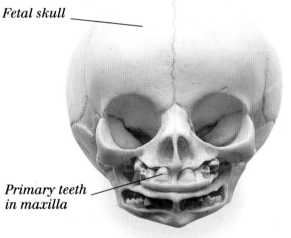

Fetal skull

Primary teeth in maxilla

DEVELOPMENT OF TEETH IN A FETUS

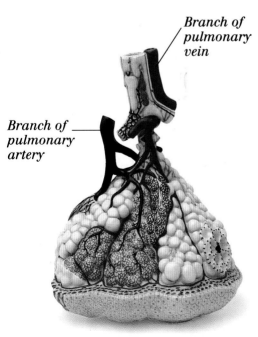

Branch of pulmonary vein

Branch of pulmonary artery

BRONCHIOLE WITH LOBULE

The human body

ALTHOUGH THERE IS enormous variation between the external appearances of humans, all bodies contain the same basic features.  The outward form of the human body depends on the size of the skeleton, the shape of the muscles,the thickness of the fat layer beneath the skin, the elasticity or sagginess of the skin, and the person's age and gender. Males tend to be taller than females, with broader shoulders, more body hair, and a different pattern of fat deposits underthe skin; the female body tends to be less muscular and has a shallower and wider pelvis to allow for childbirth.

Ear

Nape of neck

Shoulder

Scapula
(shoulder blade)

Upper arm

Back

Elbow

Loin

Waist

Forearm

Natal
cleft

Buttock

Gluteal fold

Arm

Hand

Popliteal fossa

Leg

Calf

Foot

Heel

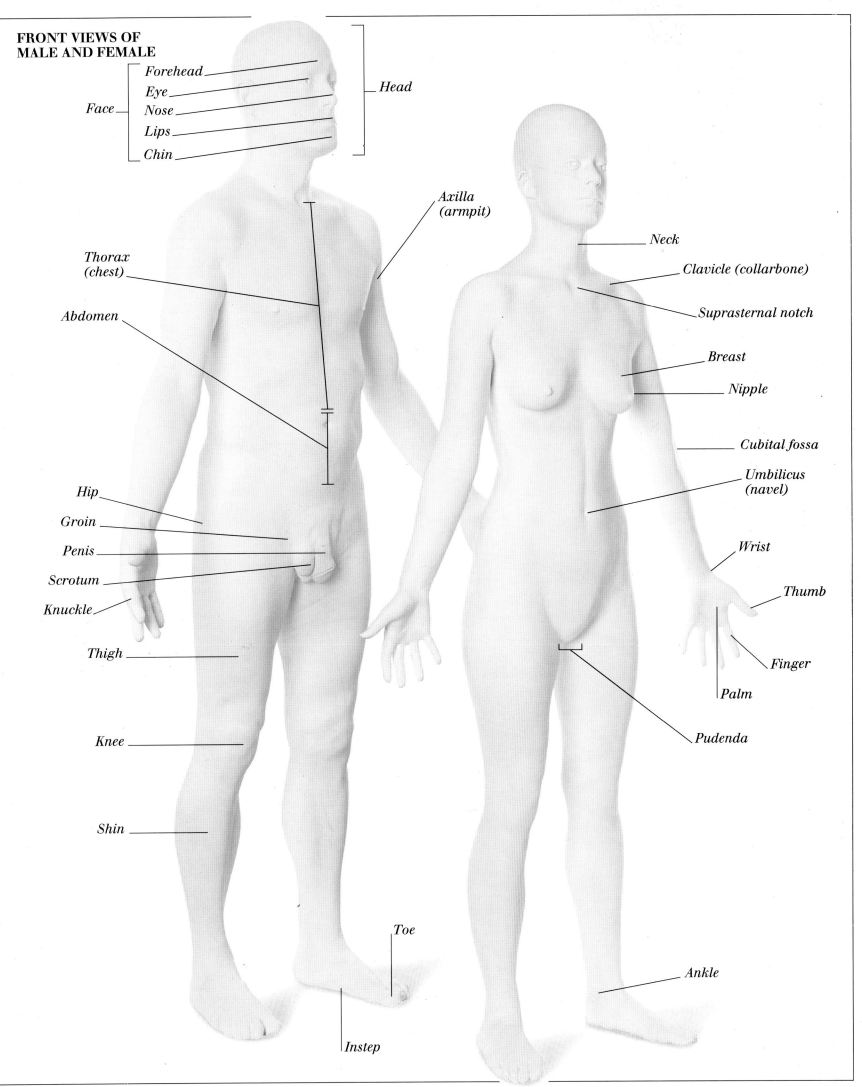

FRONT VIEWS OF MALE AND FEMALE

Forehead
Eye
Face
Nose
Lips
Chin

Head

Axilla (armpit)

Thorax (chest)

Neck

Clavicle (collarbone)

Suprasternal notch

Abdomen

Breast

Nipple

Cubital fossa

Umbilicus (navel)

Hip

Groin

Penis

Wrist

Scrotum

Thumb

Knuckle

Finger

Thigh

Palm

Knee

Pudenda

Shin

Toe

Ankle

Instep

7

Head

In a newborn baby, the head accounts for one-quarter of the total body length; by adulthood, the proportion has reduced to one-eighth. Contained in the head are the body's main sense organs: eyes, ears, olfactory nerves that detect smells, and the taste buds of the tongue. Signals from these organs pass to the body's great coordination center: the brain, housed in the protective, bony dome of the skull. Hair on the head insulates against heat loss, and adult males also grow thick facial hair. The face has three important openings: two nostrils through which air passes, and the mouth, which takes in nourishment and helps form speech. Although all heads are basically similar, differences in the size, shape, and color of features produce an infinite variety of appearances.

SIDE VIEW OF EXTERNAL FEATURES OF HEAD

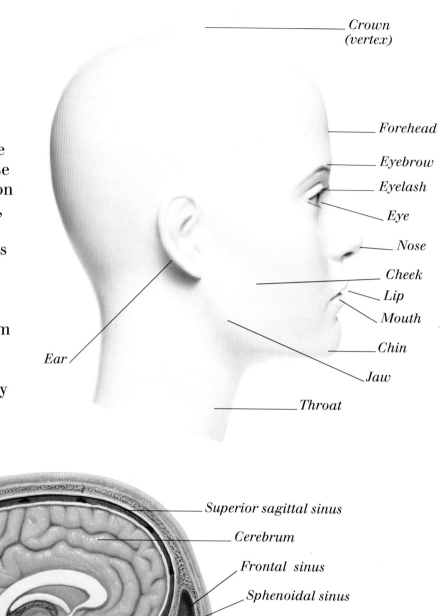

Crown (vertex)

Forehead

Eyebrow

Eyelash

Eye

Nose

Cheek

Lip

Mouth

Chin

Jaw

Throat

Ear

SECTION THROUGH HEAD

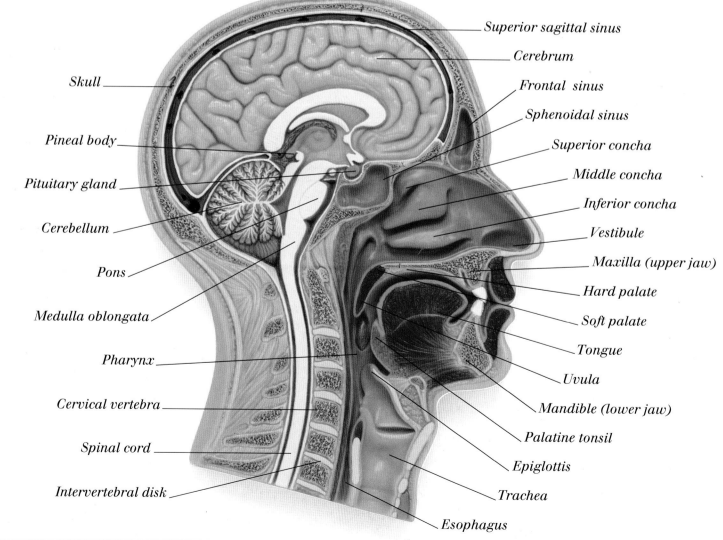

Skull

Pineal body

Pituitary gland

Cerebellum

Pons

Medulla oblongata

Pharynx

Cervical vertebra

Spinal cord

Intervertebral disk

Superior sagittal sinus

Cerebrum

Frontal sinus

Sphenoidal sinus

Superior concha

Middle concha

Inferior concha

Vestibule

Maxilla (upper jaw)

Hard palate

Soft palate

Tongue

Uvula

Mandible (lower jaw)

Palatine tonsil

Epiglottis

Trachea

Esophagus

**FRONT VIEW OF EXTERNAL
FEATURES OF HEAD**

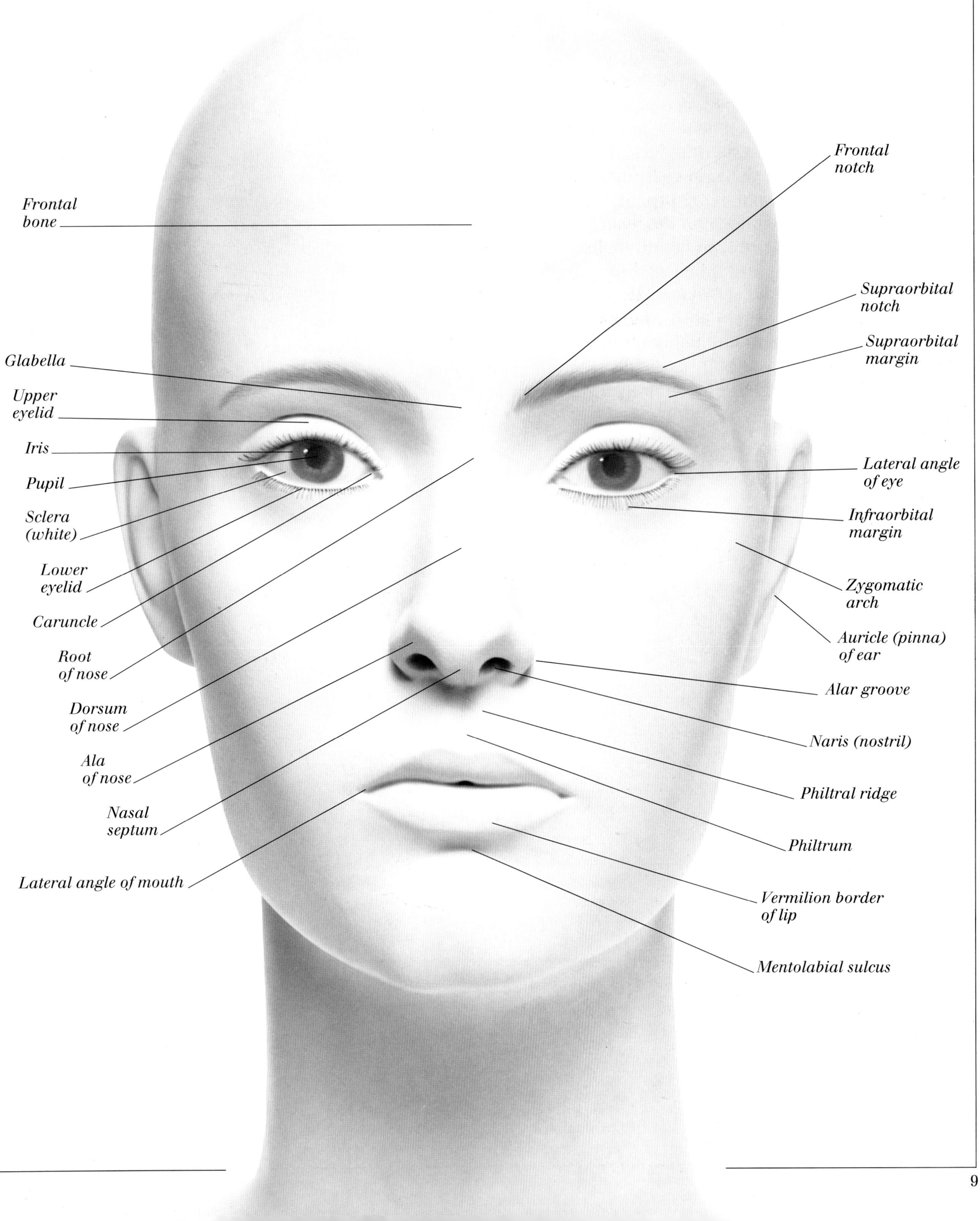

Frontal
notch

Frontal
bone

Supraorbital
notch

Supraorbital
margin

Glabella

Upper
eyelid

Iris

Pupil

Lateral angle
of eye

Sclera
(white)

Infraorbital
margin

Lower
eyelid

Zygomatic
arch

Caruncle

Auricle (pinna)
of ear

Root
of nose

Alar groove

Dorsum
of nose

Naris (nostril)

Ala
of nose

Philtral ridge

Nasal
septum

Philtrum

Lateral angle of mouth

Vermilion border
of lip

Mentolabial sulcus

Body organs

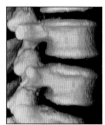

ALL THE VITAL BODY ORGANS except for the brain are enclosed within the trunk or torso (the body apart from the head and limbs). The trunk contains two large cavities separated by a muscular sheet called the diaphragm. The upper cavity, known as the thorax or chest cavity, contains the heart and lungs. The lower cavity, called the abdominal cavity, contains the stomach, intestines, liver, and pancreas, which all play a role in digesting food. Also within the trunk are the kidneys and bladder, which are part of the urinary system, and the reproductive organs, which hold the seeds of new human life. Modern imaging techniques, such as contrast X-rays and different types of scans, make it possible to see and study body organs without the need to cut through their protective coverings of skin, fat, muscle, and bone.

MAJOR INTERNAL STRUCTURES

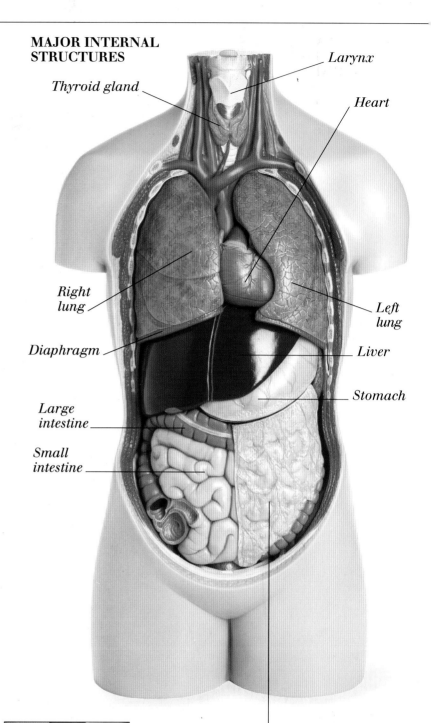

Thyroid gland

Larynx

Heart

Right lung

Left lung

Diaphragm

Liver

Large intestine

Stomach

Small intestine

Greater omentum

IMAGING THE BODY

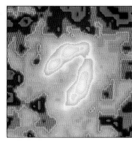

SCINTIGRAM OF HEART CHAMBERS

ANGIOGRAM OF RIGHT LUNG

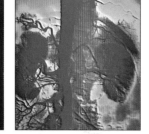

CONTRAST X-RAY OF GALLBLADDER

SCINTIGRAM OF NERVOUS SYSTEM

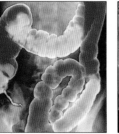

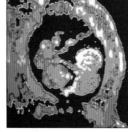

DOUBLE CONTRAST X-RAY OF COLON

ULTRASOUND SCAN OF TWINS IN UTERUS

ANGIOGRAM OF KIDNEYS

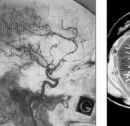

SCINTIGRAM OF NERVOUS SYSTEM

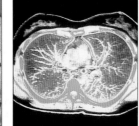

ANGIOGRAM OF ARTERIES OF HEAD

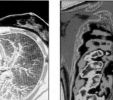

CT SCAN THROUGH FEMALE CHEST

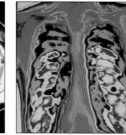

THERMOGRAM OF CHEST REGION

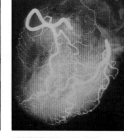

ANGIOGRAM OF ARTERIES OF HEART

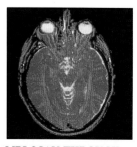

MRI SCAN THROUGH HEAD AT EYE LEVEL

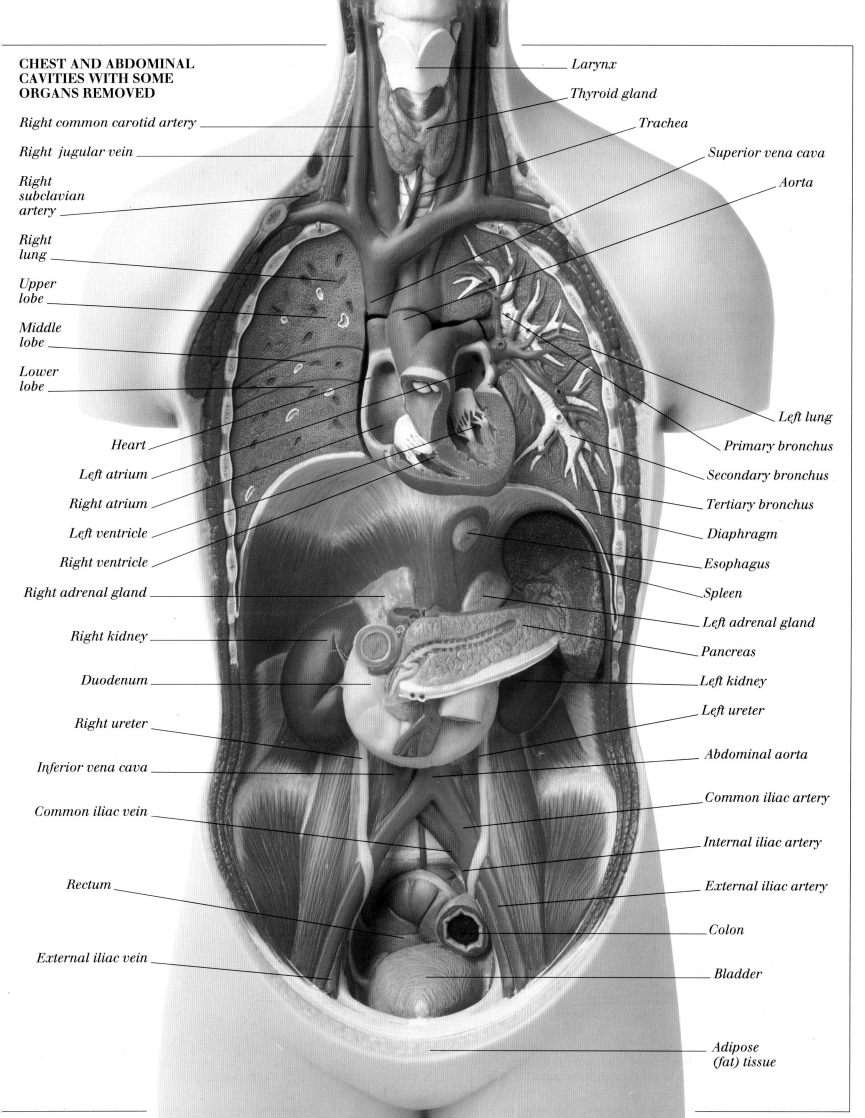

CHEST AND ABDOMINAL CAVITIES WITH SOME ORGANS REMOVED

Right common carotid artery

Right jugular vein

Right subclavian artery

Right lung

Upper lobe

Middle lobe

Lower lobe

Heart

Left atrium

Right atrium

Left ventricle

Right ventricle

Right adrenal gland

Right kidney

Duodenum

Right ureter

Inferior vena cava

Common iliac vein

Rectum

External iliac vein

Larynx

Thyroid gland

Trachea

Superior vena cava

Aorta

Left lung

Primary bronchus

Secondary bronchus

Tertiary bronchus

Diaphragm

Esophagus

Spleen

Left adrenal gland

Pancreas

Left kidney

Left ureter

Abdominal aorta

Common iliac artery

Internal iliac artery

External iliac artery

Colon

Bladder

Adipose (fat) tissue

11

Body cells

EVERYONE IS MADE UP OF BILLIONS OF CELLS, which are the basic structural units of the body. Bones, muscles, nerves, skin, blood, and all other body tissues are formed from different types of cells. Each cell has a specific function but works with other types of cells to perform the enormous number of tasks needed to sustain life. Most body cells have a similar basic structure. Each cell has an outer layer (called the cell membrane) and contains a fluid material (cytoplasm). Within the cytoplasm are many specialized structures (organelles). The most important organelle is the nucleus, which contains vital genetic material and acts as the cell's control center.

Microvillus

Adenine

Thymine

Vacuole

Nucleolus

Nuclear membrane

Cytosine

Guanine

Phosphate/sugar band

Smooth endoplasmic reticulum

THE DOUBLE HELIX
Diagrammatic representation of DNA, which is structured like a spiral ladder. DNA contains all the vital genetic information and instruction codes necessary for the maintenance and continuation of life.

Secretory vesicle

Nucleoplasm

GENERALIZED HUMAN CELL

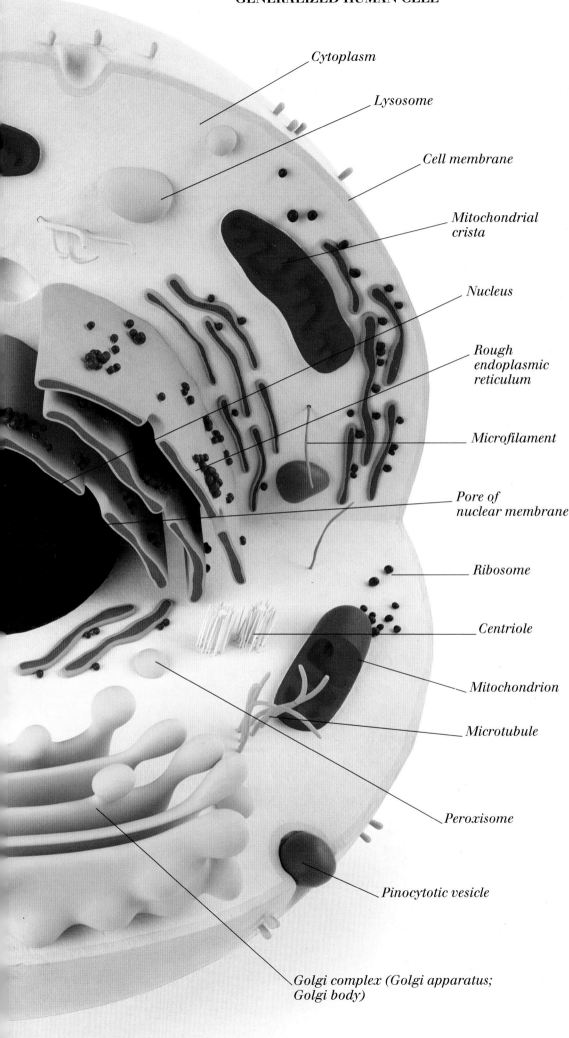

Cytoplasm

Lysosome

Cell membrane

Mitochondrial crista

Nucleus

Rough endoplasmic reticulum

Microfilament

Pore of nuclear membrane

Ribosome

Centriole

Mitochondrion

Microtubule

Peroxisome

Pinocytotic vesicle

Golgi complex (Golgi apparatus; Golgi body)

TYPES OF CELLS

BONE-FORMING CELL

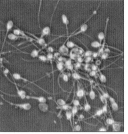

NERVE CELLS IN SPINAL CORD

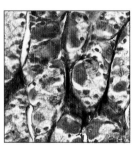

SPERM CELLS IN SEMEN

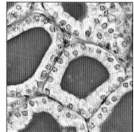

SECRETORY THYROID GLAND CELLS

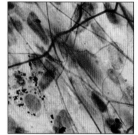

ACID-SECRETING STOMACH CELLS

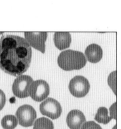

CONNECTIVE TISSUE CELLS

MUCUS-SECRETING DUODENAL CELLS

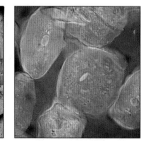

RED AND TWO WHITE BLOOD CELLS

FAT CELLS IN ADIPOSE TISSUE

EPITHELIAL CELLS IN CHEEK

Skeleton

THE SKELETON IS A MOBILE FRAMEWORK made up of 206 bones, approximately half of which are in the hands and feet. Although individual bones are rigid, the skeleton as a whole is remarkably flexible and allows the human body a huge range of movement. The skeleton serves as an anchorage for the skeletal muscles, and as a protective cage for the body's internal organs. Female bones are usually smaller and lighter than male bones, and the female pelvis is shallower and has a wider cavity.

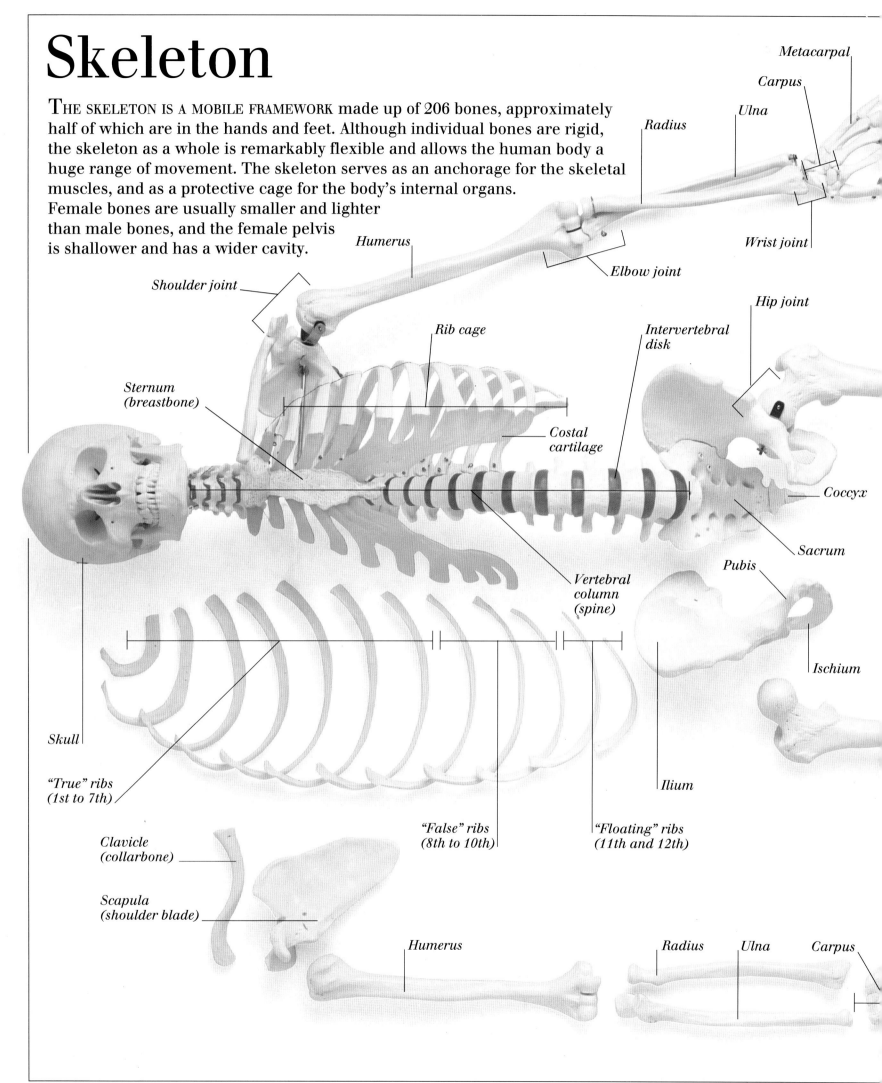

Metacarpal

Carpus

Ulna

Radius

Humerus

Shoulder joint

Elbow joint

Wrist joint

Hip joint

Rib cage

Intervertebral disk

Sternum (breastbone)

Costal cartilage

Coccyx

Vertebral column (spine)

Sacrum

Pubis

Ischium

Skull

Ilium

"True" ribs (1st to 7th)

"False" ribs (8th to 10th)

"Floating" ribs (11th and 12th)

Clavicle (collarbone)

Scapula (shoulder blade)

Humerus

Radius

Ulna

Carpus

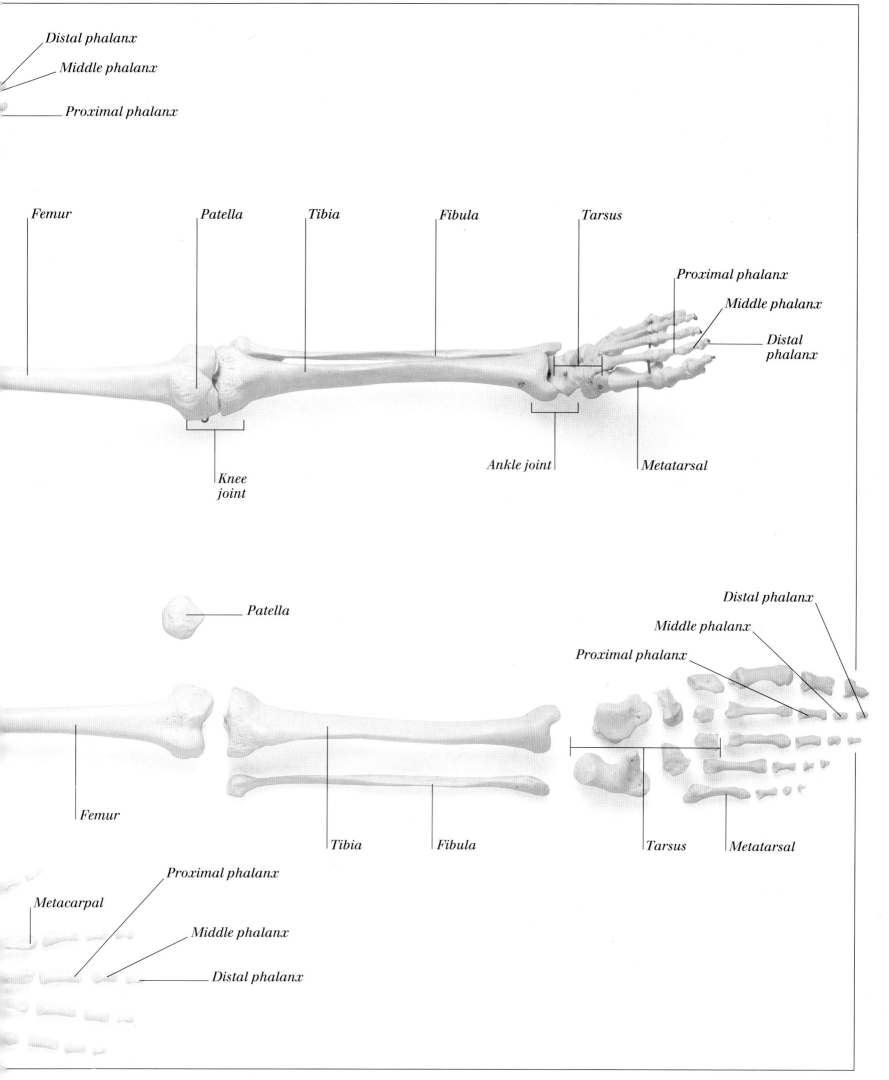

Distal phalanx

Middle phalanx

Proximal phalanx

Femur

Patella

Tibia

Fibula

Tarsus

Proximal phalanx

Middle phalanx

Distal phalanx

Knee joint

Ankle joint

Metatarsal

Patella

Distal phalanx

Middle phalanx

Proximal phalanx

Femur

Tibia

Fibula

Tarsus

Metatarsal

Proximal phalanx

Metacarpal

Middle phalanx

Distal phalanx

Skull

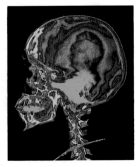

THE SKULL is the most complicated bony structure of the body—but every feature serves a purpose. Internally, the main hollow chamber of the skull has three levels that support the brain, with every bump and hollow corresponding to the shape of the brain. Underneath and toward the back of the skull is a large round hole, called the foramen magnum, through which the spinal cord passes. To the front of this are many smaller openings through which nerves, arteries, and veins pass to and from the brain. The roof of the skull is formed from four thin, curved bones that are firmly fixed together from the age of about two years. At the front of the skull are two orbits, which contain the eyeballs, and a central hole for the airway of the nose. The jawbone hinges on either side of the skull at ear level.

RIGHT SIDE VIEW OF A FETAL SKULL

Anterior fontanelle

Parietal bone

Coronal suture

Frontal bone

Nasal bone

Mental symphysis

Lambdoid suture

Occipital bone

Sphenoidal fontanelle

Mastoid fontanelle

External auditory meatus

RIGHT SIDE VIEW OF SKULL

Greater wing of sphenoid bone

Coronal suture

Frontal bone

Frontozygomatic suture

Parietal bone

Squamous suture

Supraorbital margin

Orbital cavity

VIEW OF SKULL FROM BELOW

Nasal bone

Anterior nasal spine

Maxilla (upper jaw)

Mandible (lower jaw)

Lambdoid suture

Occipital bone

Temporal bone

External auditory meatus

Mastoid process

Condyle

Coronoid process

Zygomatic bone

Mental foramen

Styloid process

Zygomatic arch

Posterior border of vomer

Concha

Mandible (lower jaw)

External occipital crest

Foramen magnum

Occipital condyle

Carotid canal

Mastoid process

Pharyngeal tubercle

Pterygoid plate

Pterygoid hamulus

Greater palatine foramen

Posterior nasal aperture

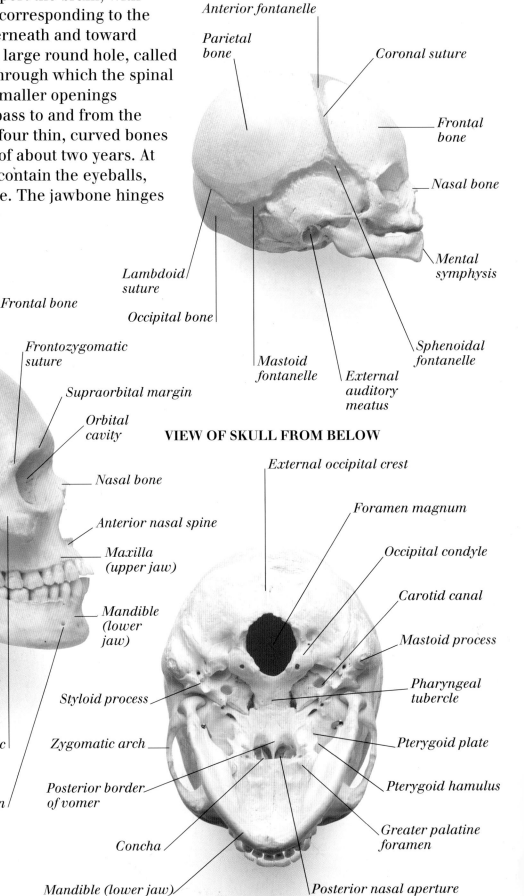

FRONT VIEW OF SKULL

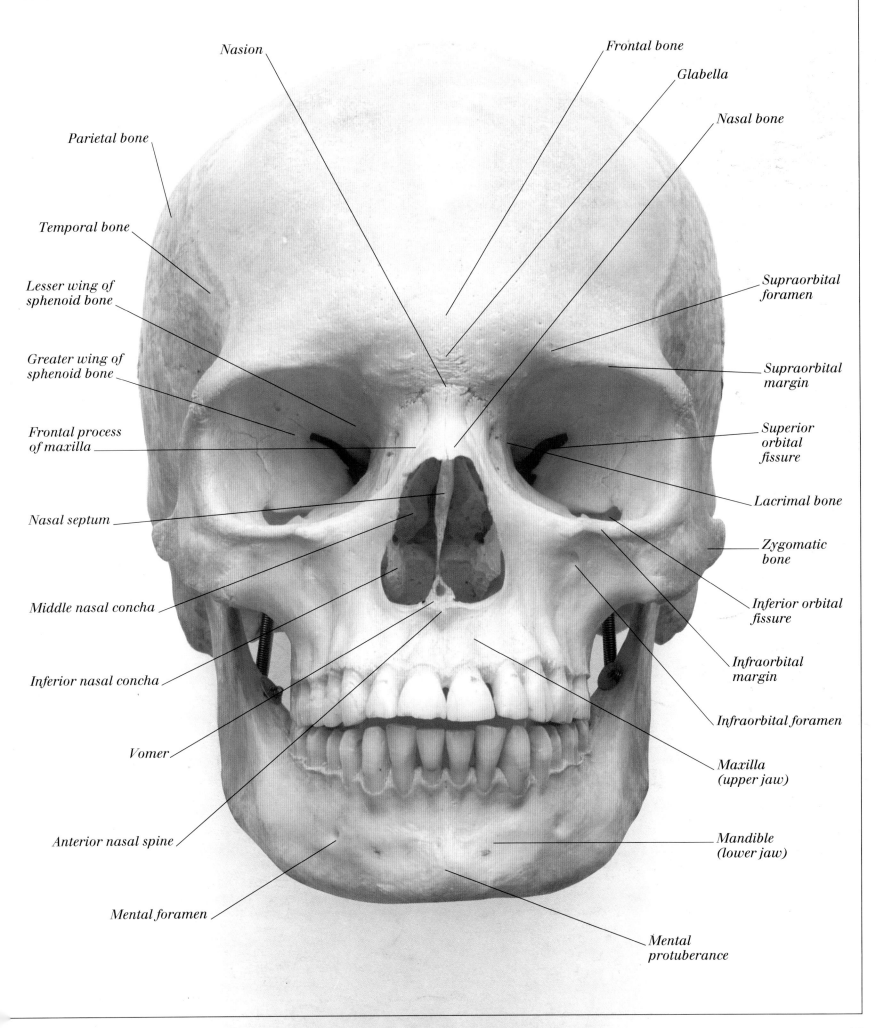

Nasion

Frontal bone

Glabella

Nasal bone

Parietal bone

Temporal bone

Supraorbital
foramen

Lesser wing of
sphenoid bone

Supraorbital
margin

Greater wing of
sphenoid bone

Superior
orbital
fissure

Frontal process
of maxilla

Lacrimal bone

Nasal septum

Zygomatic
bone

Middle nasal concha

Inferior orbital
fissure

Inferior nasal concha

Infraorbital
margin

Vomer

Infraorbital foramen

Maxilla
(upper jaw)

Anterior nasal spine

Mandible
(lower jaw)

Mental foramen

Mental
protuberance

17

Spine

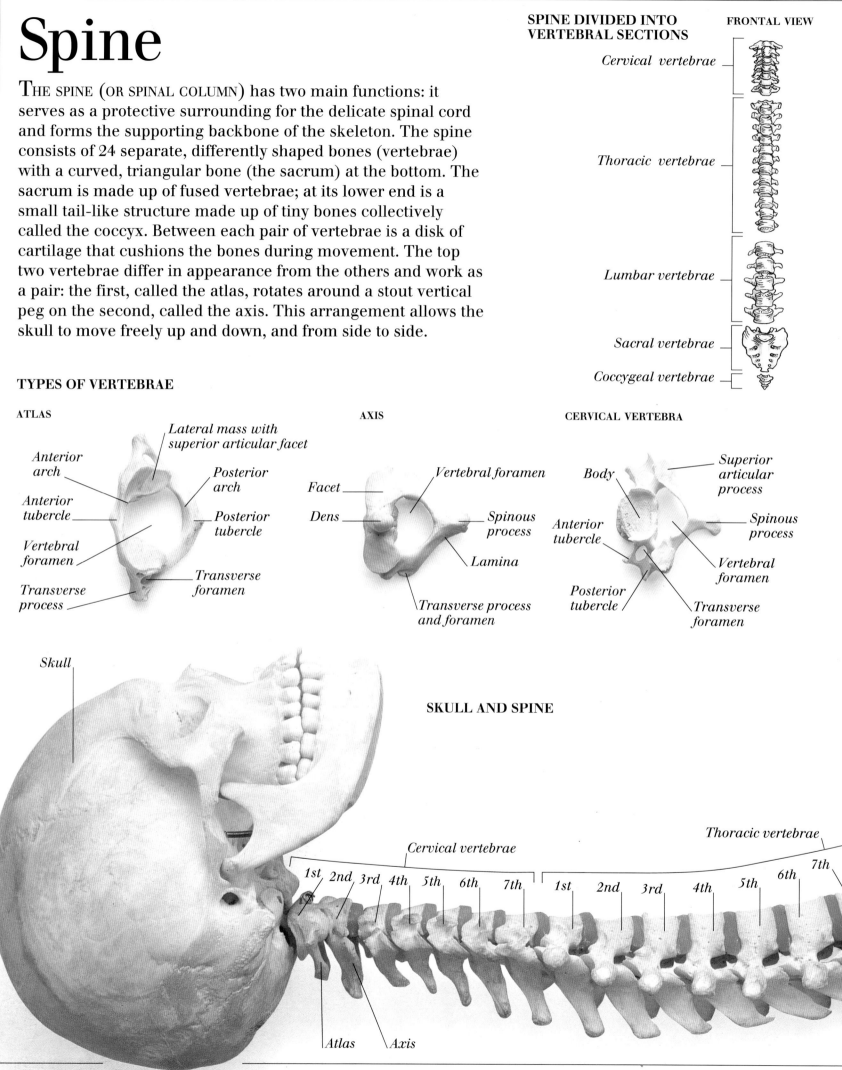

The spine (or spinal column) has two main functions: it serves as a protective surrounding for the delicate spinal cord and forms the supporting backbone of the skeleton. The spine consists of 24 separate, differently shaped bones (vertebrae) with a curved, triangular bone (the sacrum) at the bottom. The sacrum is made up of fused vertebrae; at its lower end is a small tail-like structure made up of tiny bones collectively called the coccyx. Between each pair of vertebrae is a disk of cartilage that cushions the bones during movement. The top two vertebrae differ in appearance from the others and work as a pair: the first, called the atlas, rotates around a stout vertical peg on the second, called the axis. This arrangement allows the skull to move freely up and down, and from side to side.

SPINE DIVIDED INTO VERTEBRAL SECTIONS

FRONTAL VIEW

- Cervical vertebrae
- Thoracic vertebrae
- Lumbar vertebrae
- Sacral vertebrae
- Coccygeal vertebrae

TYPES OF VERTEBRAE

ATLAS

- Anterior arch
- Anterior tubercle
- Vertebral foramen
- Transverse process
- Lateral mass with superior articular facet
- Posterior arch
- Posterior tubercle
- Transverse foramen

AXIS

- Facet
- Dens
- Vertebral foramen
- Spinous process
- Lamina
- Transverse process and foramen

CERVICAL VERTEBRA

- Body
- Anterior tubercle
- Posterior tubercle
- Superior articular process
- Spinous process
- Vertebral foramen
- Transverse foramen

Skull

SKULL AND SPINE

Cervical vertebrae

1st 2nd 3rd 4th 5th 6th 7th

Thoracic vertebrae

1st 2nd 3rd 4th 5th 6th 7th

Atlas Axis

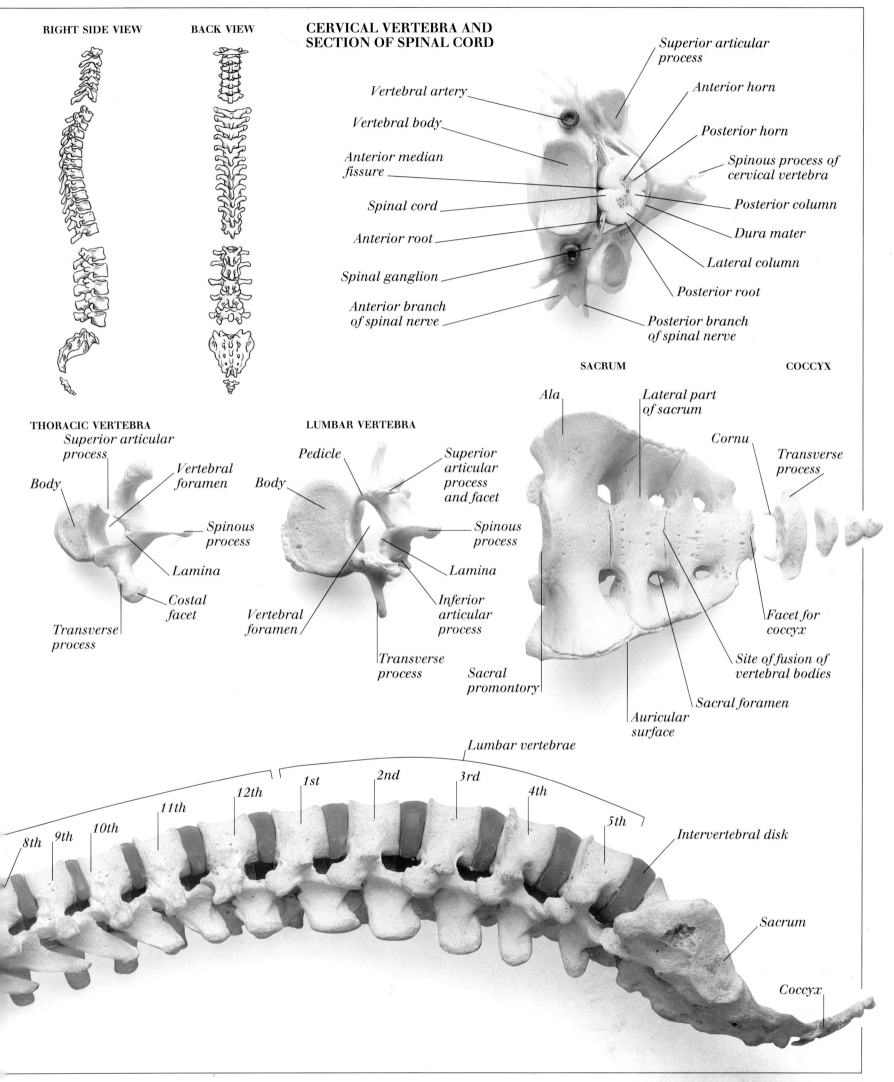

RIGHT SIDE VIEW

BACK VIEW

CERVICAL VERTEBRA AND SECTION OF SPINAL CORD

Superior articular process

Vertebral artery

Anterior horn

Vertebral body

Posterior horn

Anterior median fissure

Spinous process of cervical vertebra

Spinal cord

Posterior column

Anterior root

Dura mater

Spinal ganglion

Lateral column

Anterior branch of spinal nerve

Posterior root

Posterior branch of spinal nerve

THORACIC VERTEBRA

Superior articular process

Body

Vertebral foramen

Spinous process

Lamina

Costal facet

Transverse process

LUMBAR VERTEBRA

Pedicle

Superior articular process and facet

Body

Spinous process

Lamina

Vertebral foramen

Inferior articular process

Transverse process

Sacral promontory

SACRUM

Ala

Lateral part of sacrum

Sacral foramen

Auricular surface

Site of fusion of vertebral bodies

COCCYX

Cornu

Transverse process

Facet for coccyx

Lumbar vertebrae

8th *9th* *10th* *11th* *12th* *1st* *2nd* *3rd* *4th* *5th*

Intervertebral disk

Sacrum

Coccyx

Bones and joints

BONES FORM the body's hard, strong skeletal framework. Each bone has a hard, compact exterior surrounding a spongy, lighter interior. The long bones of the arms and legs, such as the femur (thigh bone), have a central cavity containing bone marrow. Bones are composed chiefly of calcium, phosphorus, and a fibrous substance known as collagen. Bones meet at joints, which are of several different types. For example, the hip is a ball-and-socket joint that allows the femur a wide range of movement, whereas finger joints are simple hinge joints that allow only bending and straightening. Joints are held in place by bands of tissue called ligaments. Movement of joints is facilitated by the smooth hyaline cartilage that covers the bone ends and by the synovial membrane that lines and lubricates the joint.

LIGAMENTS SURROUNDING HIP JOINT

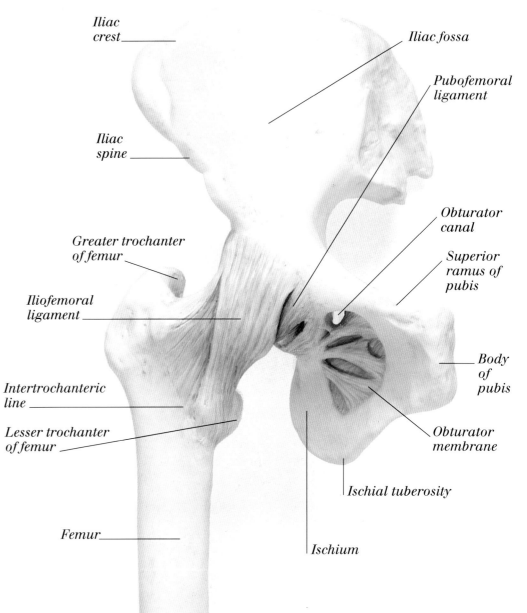

Iliac crest

Iliac fossa

Pubofemoral ligament

Iliac spine

Obturator canal

Greater trochanter of femur

Superior ramus of pubis

Iliofemoral ligament

Body of pubis

Intertrochanteric line

Obturator membrane

Lesser trochanter of femur

Ischial tuberosity

Femur

Ischium

SECTION THROUGH LEFT FEMUR

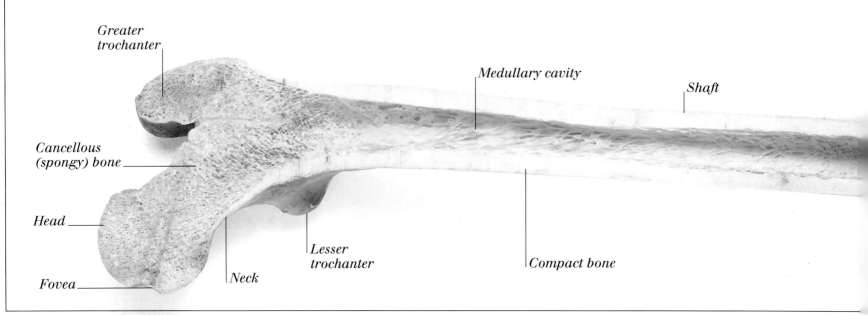

Greater trochanter

Medullary cavity

Shaft

Cancellous (spongy) bone

Head

Lesser trochanter

Compact bone

Fovea

Neck

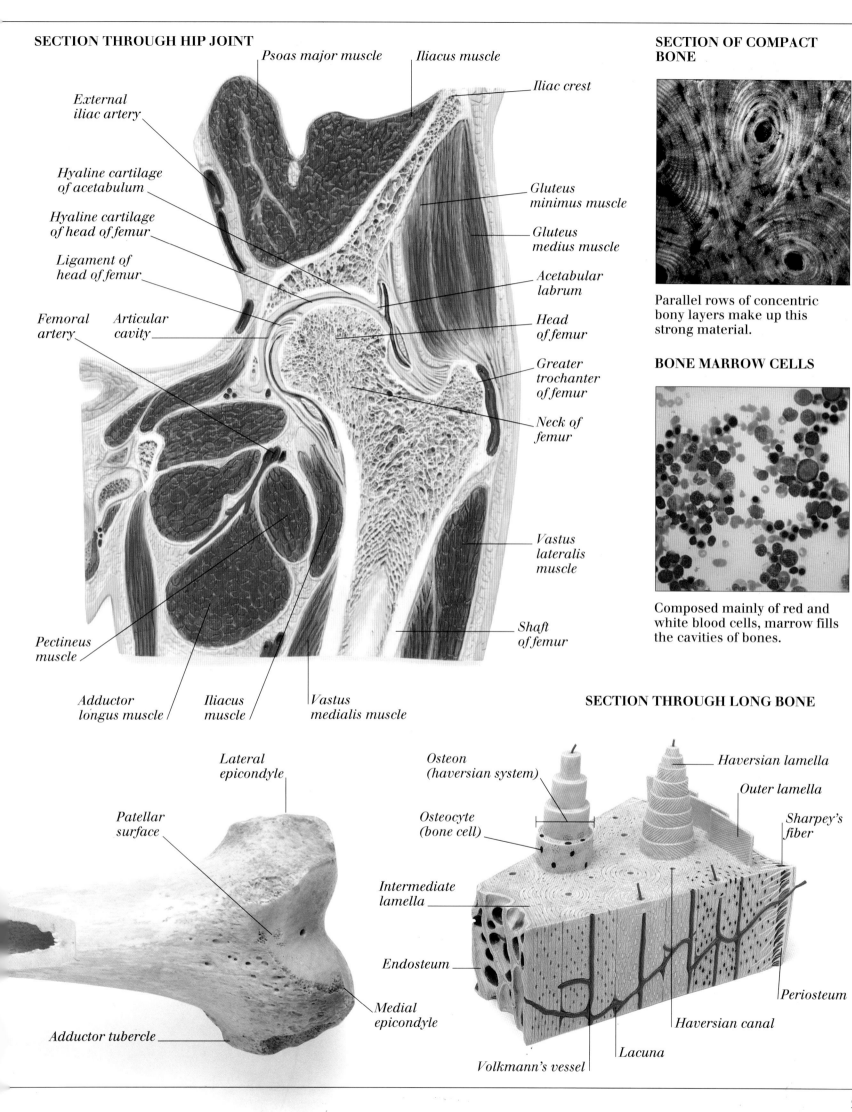

SECTION THROUGH HIP JOINT

Psoas major muscle

Iliacus muscle

Iliac crest

External iliac artery

Hyaline cartilage of acetabulum

Hyaline cartilage of head of femur

Ligament of head of femur

Femoral artery

Articular cavity

Gluteus minimus muscle

Gluteus medius muscle

Acetabular labrum

Head of femur

Greater trochanter of femur

Neck of femur

Vastus lateralis muscle

Pectineus muscle

Adductor longus muscle

Iliacus muscle

Vastus medialis muscle

Shaft of femur

SECTION OF COMPACT BONE

Parallel rows of concentric bony layers make up this strong material.

BONE MARROW CELLS

Composed mainly of red and white blood cells, marrow fills the cavities of bones.

Lateral epicondyle

Patellar surface

Adductor tubercle

Medial epicondyle

SECTION THROUGH LONG BONE

Osteon (haversian system)

Osteocyte (bone cell)

Intermediate lamella

Endosteum

Volkmann's vessel

Lacuna

Haversian canal

Haversian lamella

Outer lamella

Sharpey's fiber

Periosteum

Muscles 1

THERE ARE THREE MAIN TYPES OF MUSCLE: skeletal muscle (also called voluntary muscle because it can be consciously controlled); smooth muscle (also called involuntary muscle because it is not under voluntary control); and the specialized muscle tissue of the heart. Humans have more than 600 skeletal muscles, which differ in size and shape according to the jobs they do. Skeletal muscles are attached either directly or indirectly (via tendons) to bones, and work in opposing pairs (one muscle in the pair contracts while the other relaxes) to produce body movements as diverse as walking, threading a needle, and an array of facial expressions. Smooth muscles occur in the walls of internal body organs and perform actions such as forcing food through the intestines, contracting the uterus (womb) in childbirth, and pumping blood through the blood vessels.

SOME OTHER MUSCLES IN THE BODY

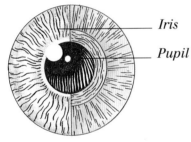

Iris

Pupil

IRIS
The muscle fibers contract and dilate (expand) to alter pupil size.

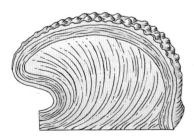

TONGUE
Interlacing layers of muscle allow great mobility.

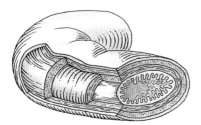

ILEUM
Opposing muscle layers transport semidigested food.

SUPERFICIAL SKELETAL MUSCLES

FRONT VIEW

Brachioradialis

Flexors of forearm

Brachialis

Frontalis

Orbicularis oculi

Temporalis

Sternocleidomastoid

Trapezius

Pectoralis major

Deltoid

Serratus anterior

Rectus abdominis

Biceps brachii

Linea alba

External oblique

Tensor fasciae latae

Iliopsoas

Pectineus

Adductor longus

Vastus lateralis

Rectus femoris

Gracilis

Sartorius

Vastus medialis

Gastrocnemius

Tibialis anterior

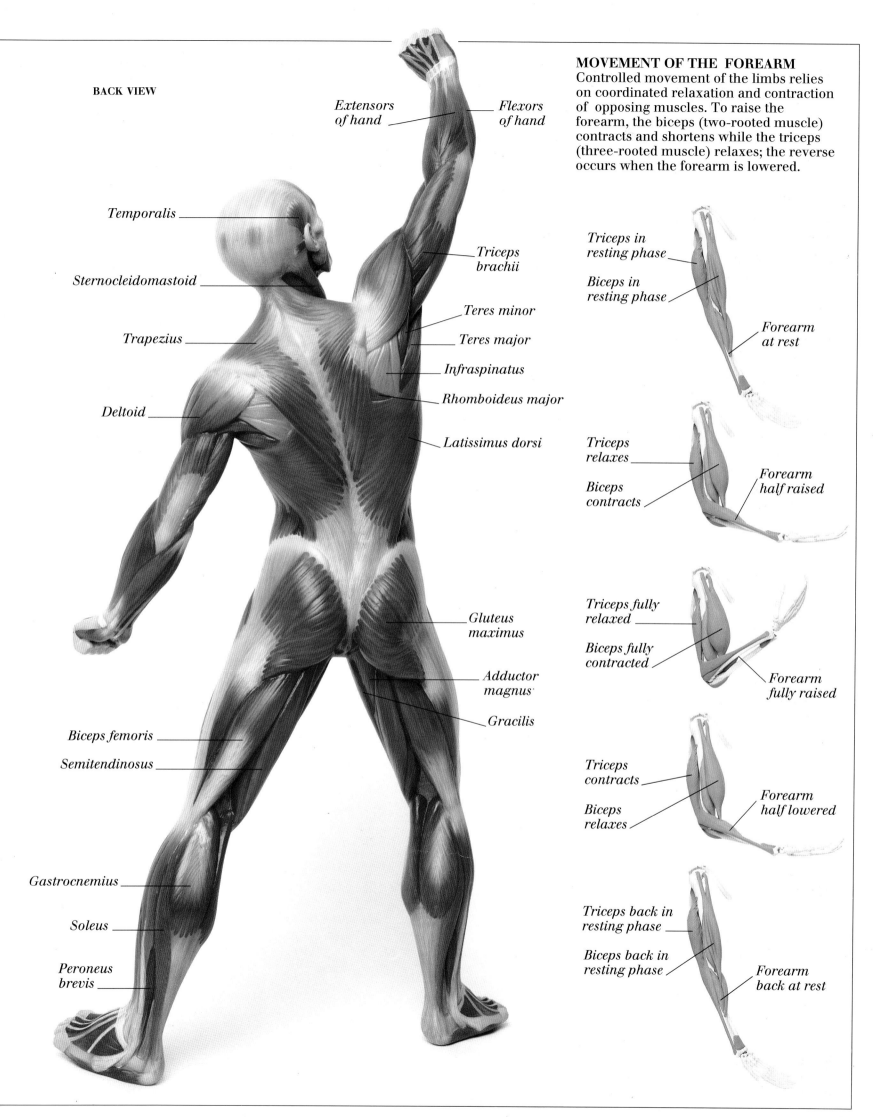

BACK VIEW

Extensors of hand

Flexors of hand

Temporalis

Sternocleidomastoid

Trapezius

Deltoid

Triceps brachii

Teres minor

Teres major

Infraspinatus

Rhomboideus major

Latissimus dorsi

Gluteus maximus

Adductor magnus

Gracilis

Biceps femoris

Semitendinosus

Gastrocnemius

Soleus

Peroneus brevis

MOVEMENT OF THE FOREARM
Controlled movement of the limbs relies on coordinated relaxation and contraction of opposing muscles. To raise the forearm, the biceps (two-rooted muscle) contracts and shortens while the triceps (three-rooted muscle) relaxes; the reverse occurs when the forearm is lowered.

Triceps in resting phase

Biceps in resting phase

Forearm at rest

Triceps relaxes

Biceps contracts

Forearm half raised

Triceps fully relaxed

Biceps fully contracted

Forearm fully raised

Triceps contracts

Biceps relaxes

Forearm half lowered

Triceps back in resting phase

Biceps back in resting phase

Forearm back at rest

23

Muscles 2

SKELETAL MUSCLE FIBER

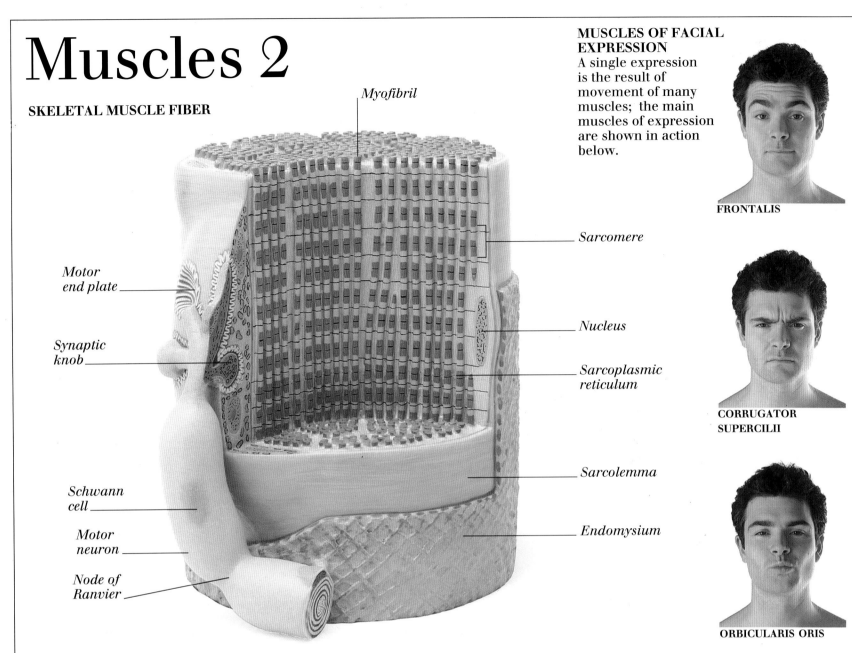

Myofibril

Sarcomere

Motor end plate

Nucleus

Synaptic knob

Sarcoplasmic reticulum

Sarcolemma

Schwann cell

Endomysium

Motor neuron

Node of Ranvier

MUSCLES OF FACIAL EXPRESSION
A single expression is the result of movement of many muscles; the main muscles of expression are shown in action below.

FRONTALIS

CORRUGATOR SUPERCILII

ORBICULARIS ORIS

ZYGOMATICUS MAJOR

DEPRESSOR ANGULI ORIS

TYPES OF MUSCLE

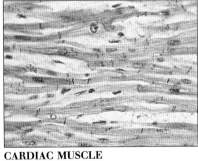

CARDIAC MUSCLE

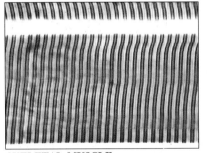

SKELETAL MUSCLE

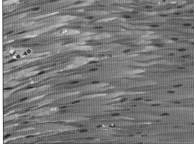

SMOOTH MUSCLE

CONTRACTION OF SKELETAL MUSCLE

RELAXED STATE

CONTRACTED STATE

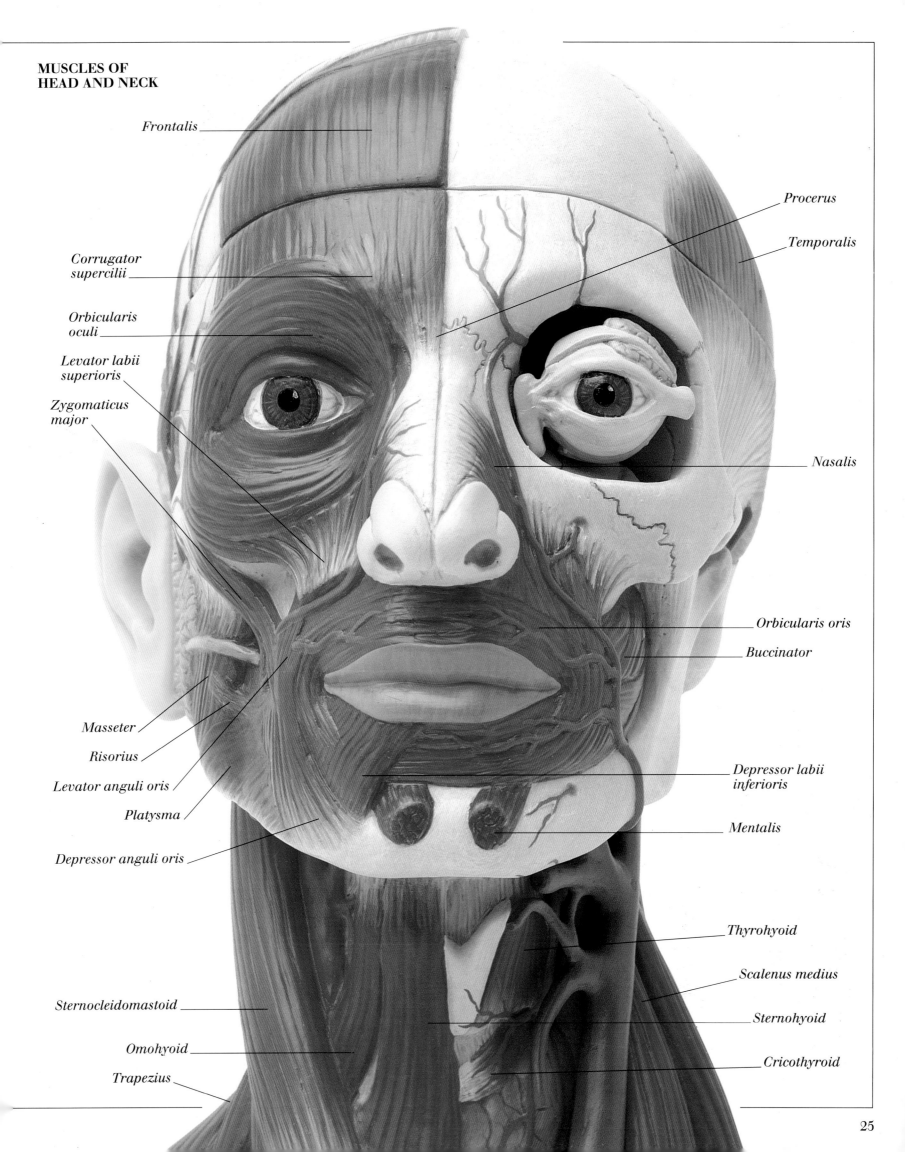

MUSCLES OF HEAD AND NECK

Frontalis

Procerus

Temporalis

Corrugator supercilii

Orbicularis oculi

Levator labii superioris

Zygomaticus major

Nasalis

Orbicularis oris

Buccinator

Depressor labii inferioris

Masseter

Risorius

Levator anguli oris

Platysma

Mentalis

Depressor anguli oris

Thyrohyoid

Scalenus medius

Sternocleidomastoid

Sternohyoid

Omohyoid

Cricothyroid

Trapezius

25

Hands

THE HUMAN HAND is an extremely versatile tool, capable of delicate manipulation as well as powerful gripping actions. The arrangement of its 27 small bones, moved by 37 skeletal muscles that are connected to the bones by tendons, allows a wide range of movements. In particular, it is our ability to bring the tips of our thumbs and fingers together, combined with the extraordinary sensitivity of our fingertips due to their rich supply of nerve endings, that gives human hands their unique dexterity.

X-RAY OF LEFT HAND OF A YOUNG CHILD

Area of ossification in phalanx

Area of ossification in metacarpal

Area of ossification in wrist

Epiphysis of ulna

Epiphysis of radius

Areas of cartilage in the wrist and at the ends of the fingerbones are the sites of growth and have still to ossify.

BONES OF HAND

Ring finger

Middle finger

Index finger

Little finger

Distal phalanx

Middle phalanx

Proximal phalanx

2nd metacarpal

3rd metacarpal

4th metacarpal

5th metacarpal

Hamate

Pisiform

Capitate

Triquetral

Lunate

Ulna

Head

Shaft

Base

Distal phalanx of thumb

Proximal phalanx of thumb

1st metacarpal

Trapezium

Trapezoid

Scaphoid

Radius

STRUCTURES UNDERLYING SKIN OF PALM OF HAND

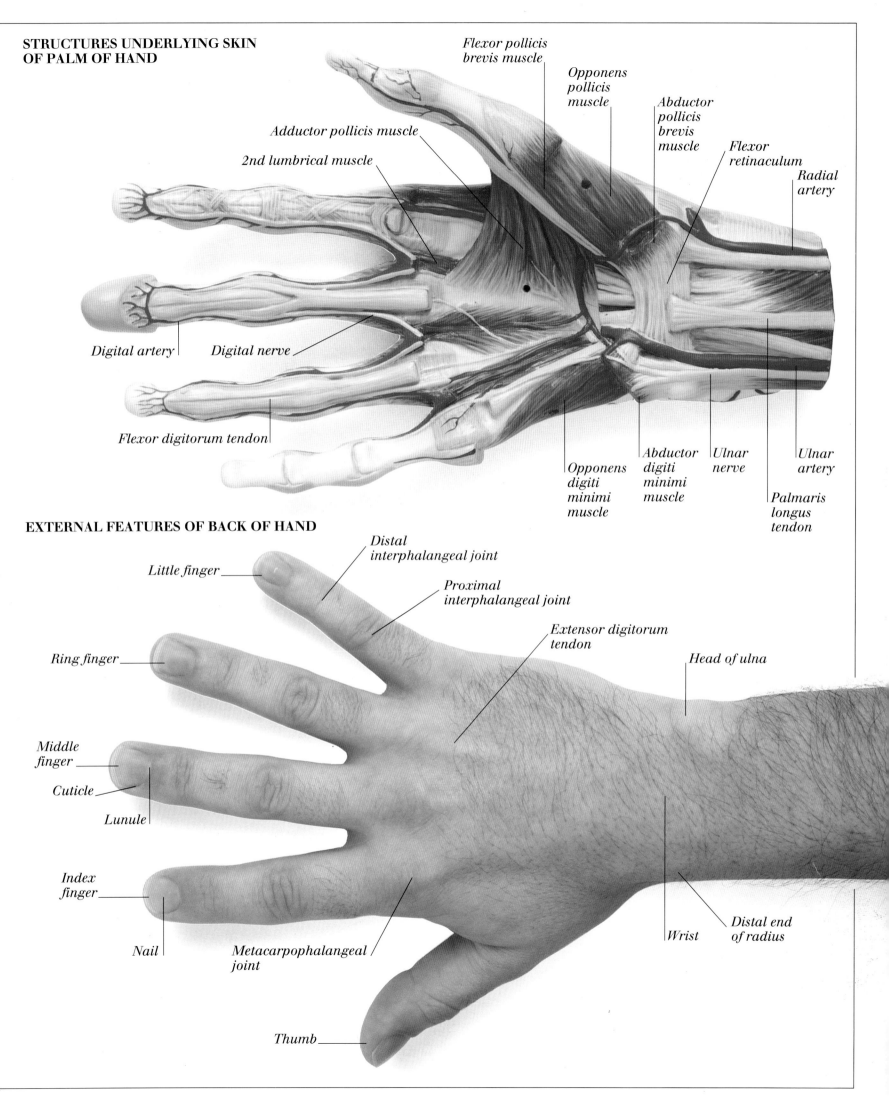

Flexor pollicis brevis muscle

Opponens pollicis muscle

Abductor pollicis brevis muscle

Flexor retinaculum

Radial artery

Adductor pollicis muscle

2nd lumbrical muscle

Digital artery

Digital nerve

Flexor digitorum tendon

Opponens digiti minimi muscle

Abductor digiti minimi muscle

Ulnar nerve

Ulnar artery

Palmaris longus tendon

EXTERNAL FEATURES OF BACK OF HAND

Distal interphalangeal joint

Little finger

Proximal interphalangeal joint

Extensor digitorum tendon

Head of ulna

Ring finger

Middle finger

Cuticle

Lunule

Index finger

Nail

Metacarpophalangeal joint

Thumb

Wrist

Distal end of radius

Feet

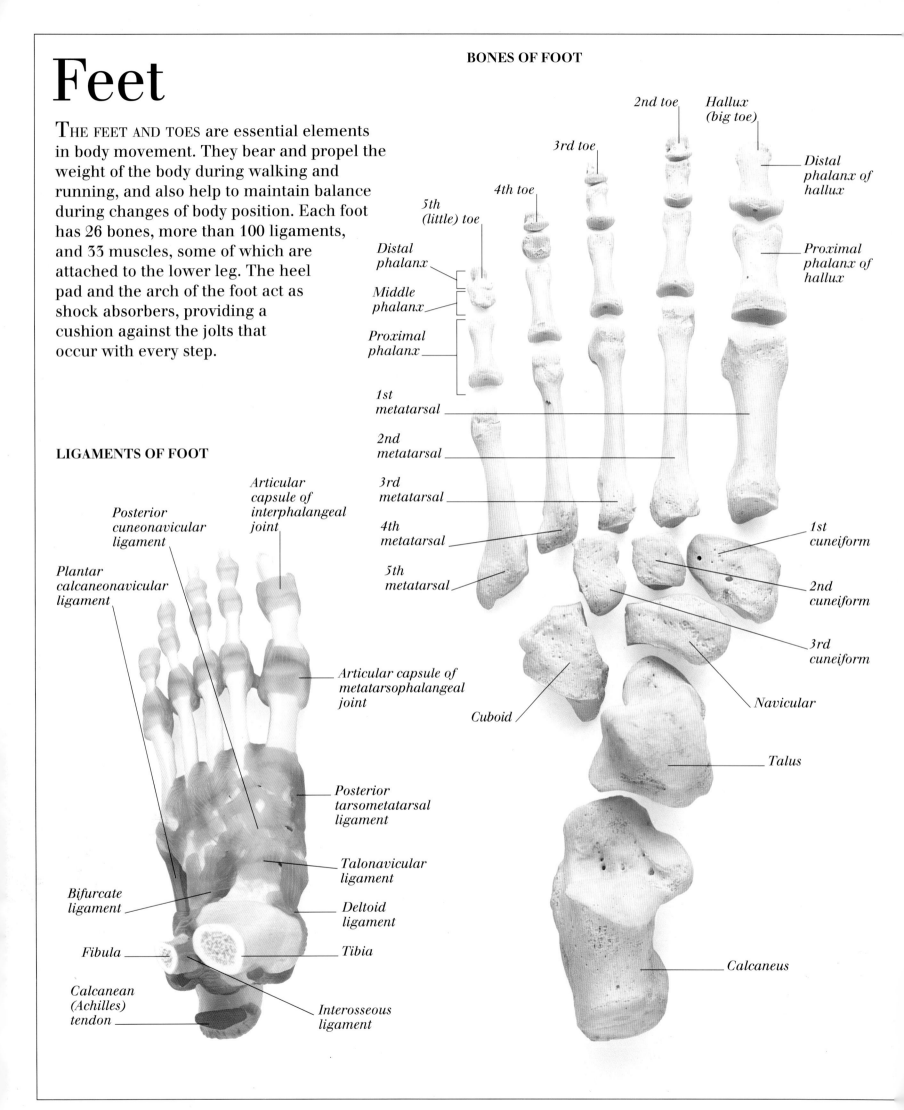

THE FEET AND TOES are essential elements in body movement. They bear and propel the weight of the body during walking and running, and also help to maintain balance during changes of body position. Each foot has 26 bones, more than 100 ligaments, and 33 muscles, some of which are attached to the lower leg. The heel pad and the arch of the foot act as shock absorbers, providing a cushion against the jolts that occur with every step.

BONES OF FOOT

2nd toe

Hallux (big toe)

3rd toe

Distal phalanx of hallux

4th toe

5th (little) toe

Proximal phalanx of hallux

Distal phalanx

Middle phalanx

Proximal phalanx

1st metatarsal

2nd metatarsal

3rd metatarsal

1st cuneiform

4th metatarsal

2nd cuneiform

5th metatarsal

3rd cuneiform

Navicular

Cuboid

Talus

Calcaneus

LIGAMENTS OF FOOT

Articular capsule of interphalangeal joint

Posterior cuneonavicular ligament

Plantar calcaneonavicular ligament

Articular capsule of metatarsophalangeal joint

Posterior tarsometatarsal ligament

Talonavicular ligament

Bifurcate ligament

Deltoid ligament

Fibula

Tibia

Calcanean (Achilles) tendon

Interosseous ligament

STRUCTURES UNDERLYING SKIN OF FOOT

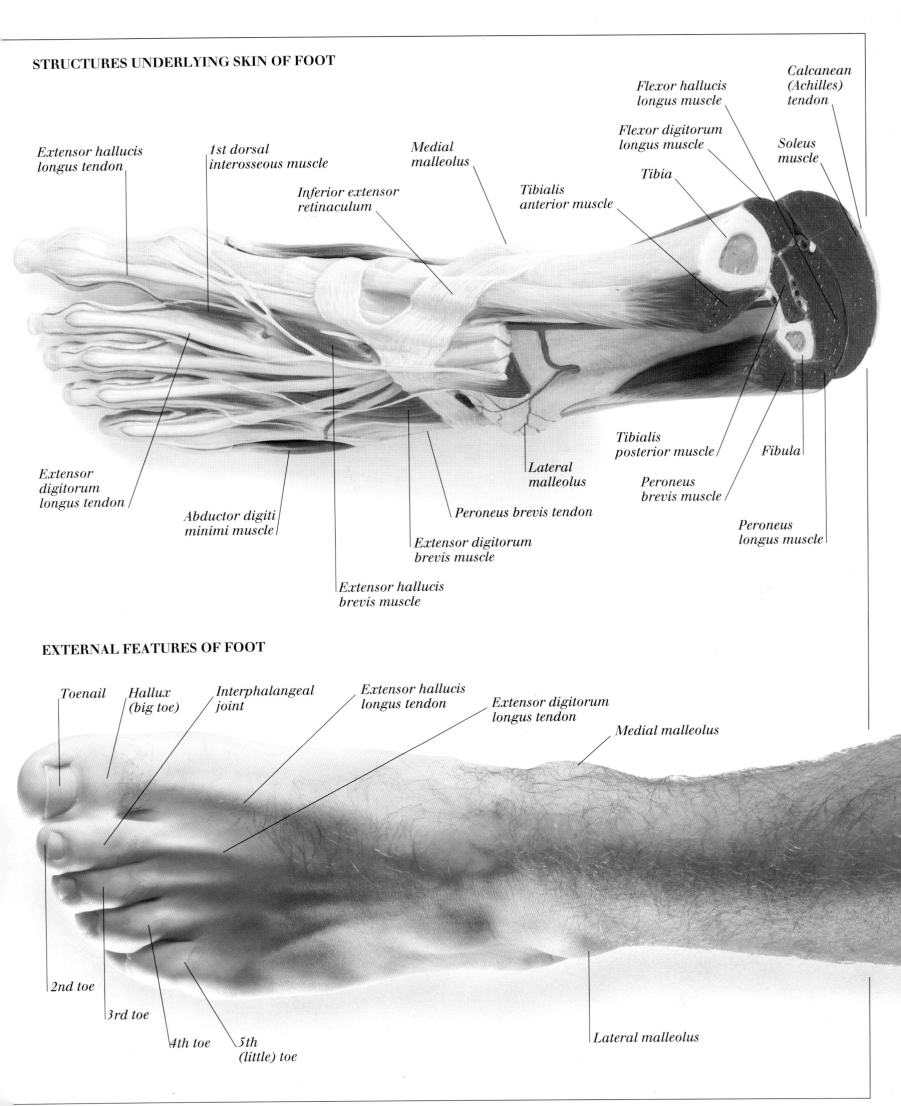

Extensor hallucis longus tendon

1st dorsal interosseous muscle

Medial malleolus

Flexor hallucis longus muscle

Calcanean (Achilles) tendon

Flexor digitorum longus muscle

Soleus muscle

Inferior extensor retinaculum

Tibialis anterior muscle

Tibia

Extensor digitorum longus tendon

Tibialis posterior muscle

Fibula

Abductor digiti minimi muscle

Lateral malleolus

Peroneus brevis muscle

Extensor digitorum brevis muscle

Peroneus brevis tendon

Peroneus longus muscle

Extensor hallucis brevis muscle

EXTERNAL FEATURES OF FOOT

Toenail

Hallux (big toe)

Interphalangeal joint

Extensor hallucis longus tendon

Extensor digitorum longus tendon

Medial malleolus

2nd toe

3rd toe

4th toe

5th (little) toe

Lateral malleolus

29

Skin and hair

SKIN IS THE BODY'S LARGEST ORGAN, a waterproof barrier that protects the internal organs against infection, injury, and harmful sun rays. The skin is also an important sensory organ and helps to control body temperature. The outer layer of the skin, known as the epidermis, is coated with keratin, a tough, horny protein that is also the chief consistituent of hair and nails. Dead cells are shed from the skin's surface and are replaced by new cells from the base of the epidermis, the region that also produces the skin pigment, melanin. The dermis contains most of the skin's living structures, and includes nerve endings, blood vessels, elastic fibers, sweat glands that cool the skin, and sebaceous glands that produce oil to keep the skin supple. Beneath the dermis lies the subcutaneous tissue (hypodermis), which is rich in fat and blood vessels. Hair shafts grow from hair follicles situated in the dermis and subcutaneous tissue. Hair grows on every part of the skin apart from the palms of the hands and soles of the feet.

SECTION OF HAIR

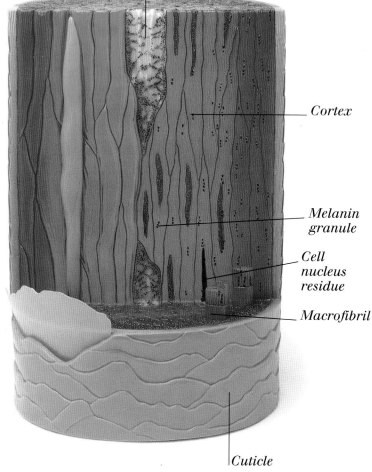

- Medulla
- Cortex
- Melanin granule
- Cell nucleus residue
- Macrofibril
- Cuticle

SECTIONS OF DIFFERENT TYPES OF SKIN

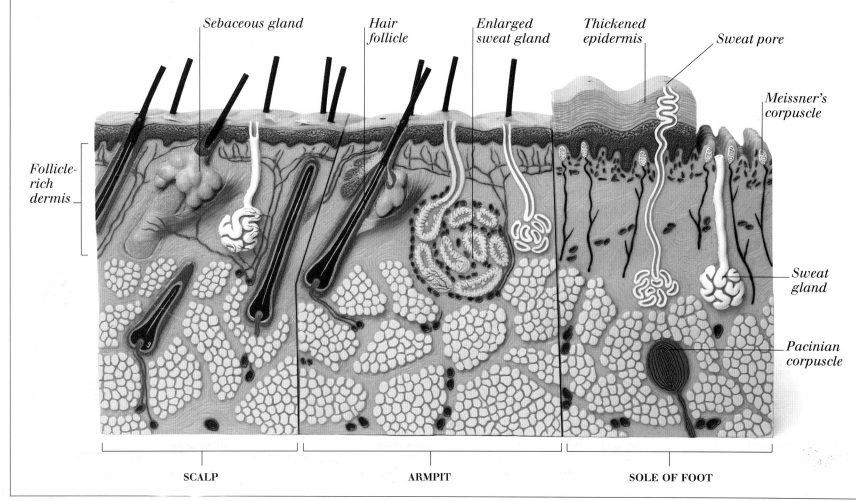

- Sebaceous gland
- Hair follicle
- Enlarged sweat gland
- Thickened epidermis
- Sweat pore
- Meissner's corpuscle
- Follicle-rich dermis
- Sweat gland
- Pacinian corpuscle

SCALP

ARMPIT

SOLE OF FOOT

SECTION OF SKIN

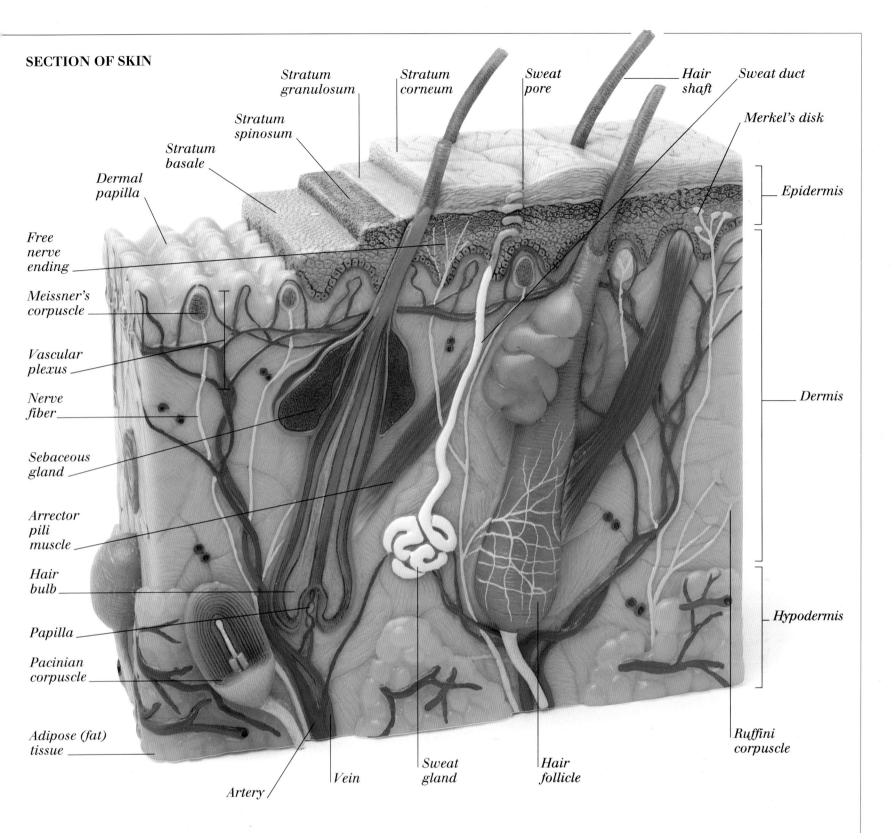

Stratum
granulosum

Stratum
spinosum

Stratum corneum

Stratum
basale

Dermal
papilla

Sweat
pore

Hair
shaft

Sweat duct

Merkel's disk

Epidermis

Free
nerve
ending

Meissner's
corpuscle

Vascular
plexus

Nerve
fiber

Sebaceous
gland

Arrector
pili
muscle

Hair
bulb

Papilla

Pacinian
corpuscle

Adipose (fat)
tissue

Dermis

Hypodermis

Ruffini
corpuscle

Artery

Vein

Sweat
gland

Hair
follicle

PHOTOMICROGRAPHS OF SKIN AND HAIR

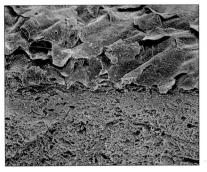

SECTION OF SKIN
The flaky cells at the skin's
surface are shed continuously.

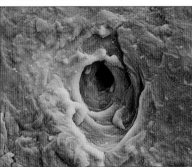

SWEAT PORE
This allows loss of fluid as part
of temperature control.

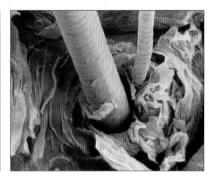

SKIN HAIR
Two hairs pushing through the
outer layer of skin.

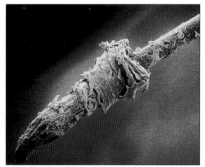

HEAD HAIR
The root and part of the shaft of
a hair from the scalp.

Brain

THE BRAIN IS THE MAJOR ORGAN of the central nervous system and the control center for all the body's voluntary and involuntary activities. It is also responsible for the complexities of thought, memory, emotion, and language. In adults, this complex organ is a mere 3 lb (1.4 kg) in weight, containing over 10 thousand million nerve cells. Three distinct regions can easily be seen—the brainstem, the cerebellum, and the large cerebrum. The brainstem controls vital body functions, such as breathing and digestion. The cerebellum's main functions are the maintenance of posture and the coordination of body movements. The cerebrum, which consists of the right and left cerebral hemispheres joined by the corpus callosum, is the site of most conscious and intelligent activities.

MRI SCAN OF TRANSVERSE SECTION THROUGH BRAIN

White matter
Skull
Scalp
Gray matter
Lateral ventricle
Longitudinal fissure
Coronal section
Sagittal section

SAGITTAL SECTION THROUGH BRAIN

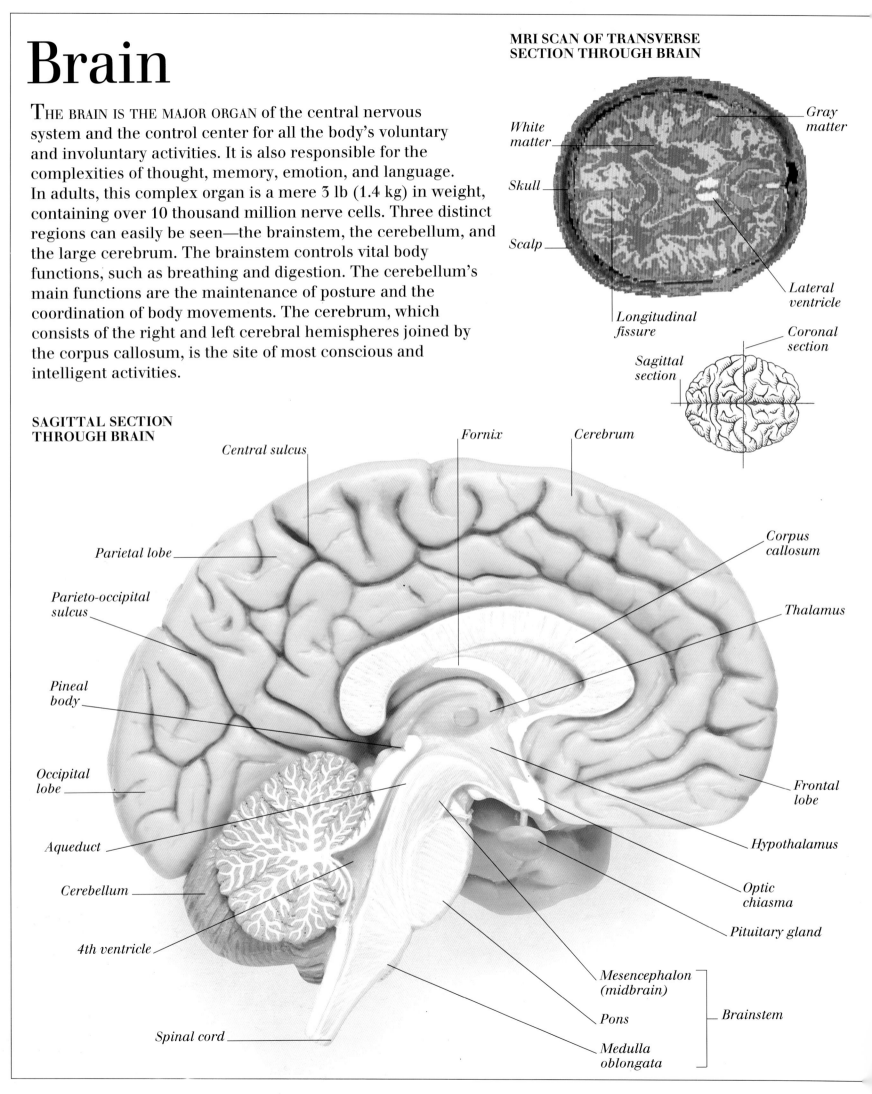

Central sulcus
Fornix
Cerebrum
Parietal lobe
Corpus callosum
Parieto-occipital sulcus
Thalamus
Pineal body
Occipital lobe
Frontal lobe
Aqueduct
Hypothalamus
Cerebellum
Optic chiasma
4th ventricle
Pituitary gland
Spinal cord
Mesencephalon (midbrain)
Pons
Brainstem
Medulla oblongata

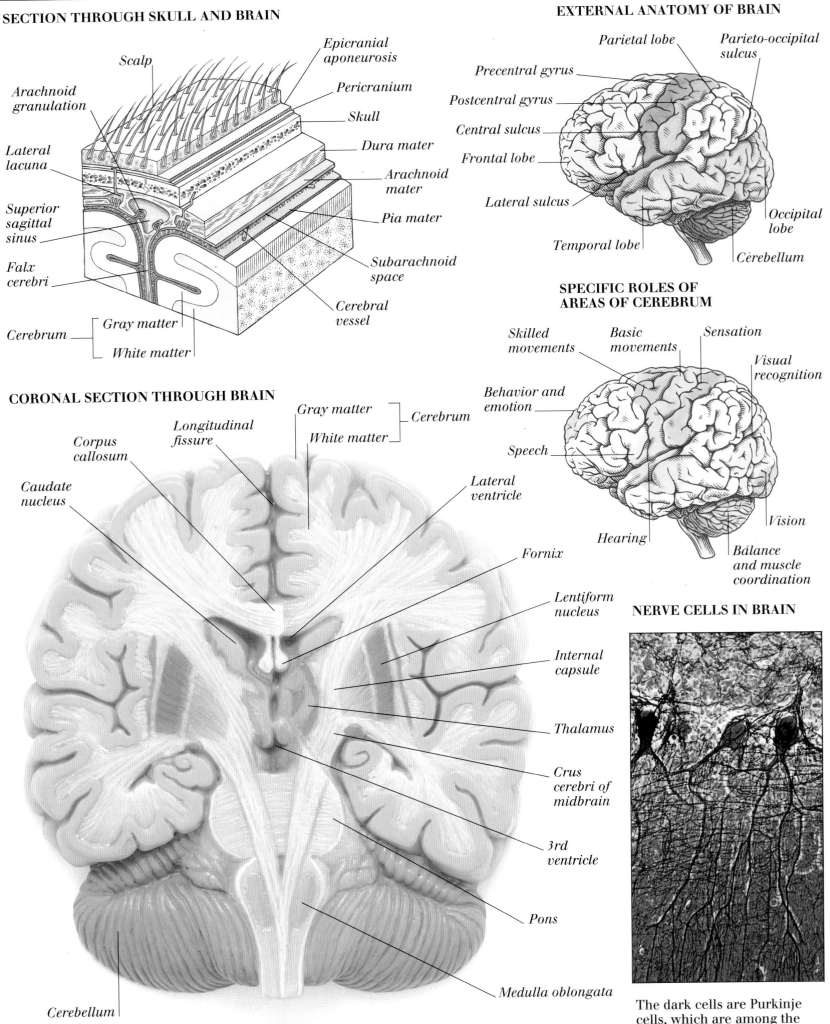

SECTION THROUGH SKULL AND BRAIN

Scalp

Epicranial aponeurosis

Arachnoid granulation

Pericranium

Skull

Lateral lacuna

Dura mater

Arachnoid mater

Superior sagittal sinus

Pia mater

Falx cerebri

Subarachnoid space

Cerebral vessel

Cerebrum { Gray matter

White matter

EXTERNAL ANATOMY OF BRAIN

Parietal lobe

Parieto-occipital sulcus

Precentral gyrus

Postcentral gyrus

Central sulcus

Frontal lobe

Lateral sulcus

Occipital lobe

Temporal lobe

Cerebellum

SPECIFIC ROLES OF AREAS OF CEREBRUM

Skilled movements

Basic movements

Sensation

Visual recognition

Behavior and emotion

Speech

Hearing

Vision

Balance and muscle coordination

CORONAL SECTION THROUGH BRAIN

Corpus callosum

Longitudinal fissure

Gray matter

White matter } Cerebrum

Caudate nucleus

Lateral ventricle

Fornix

Lentiform nucleus

Internal capsule

Thalamus

Crus cerebri of midbrain

3rd ventricle

Pons

Medulla oblongata

Cerebellum

NERVE CELLS IN BRAIN

The dark cells are Purkinje cells, which are among the largest nerve cells in the body.

Nervous system

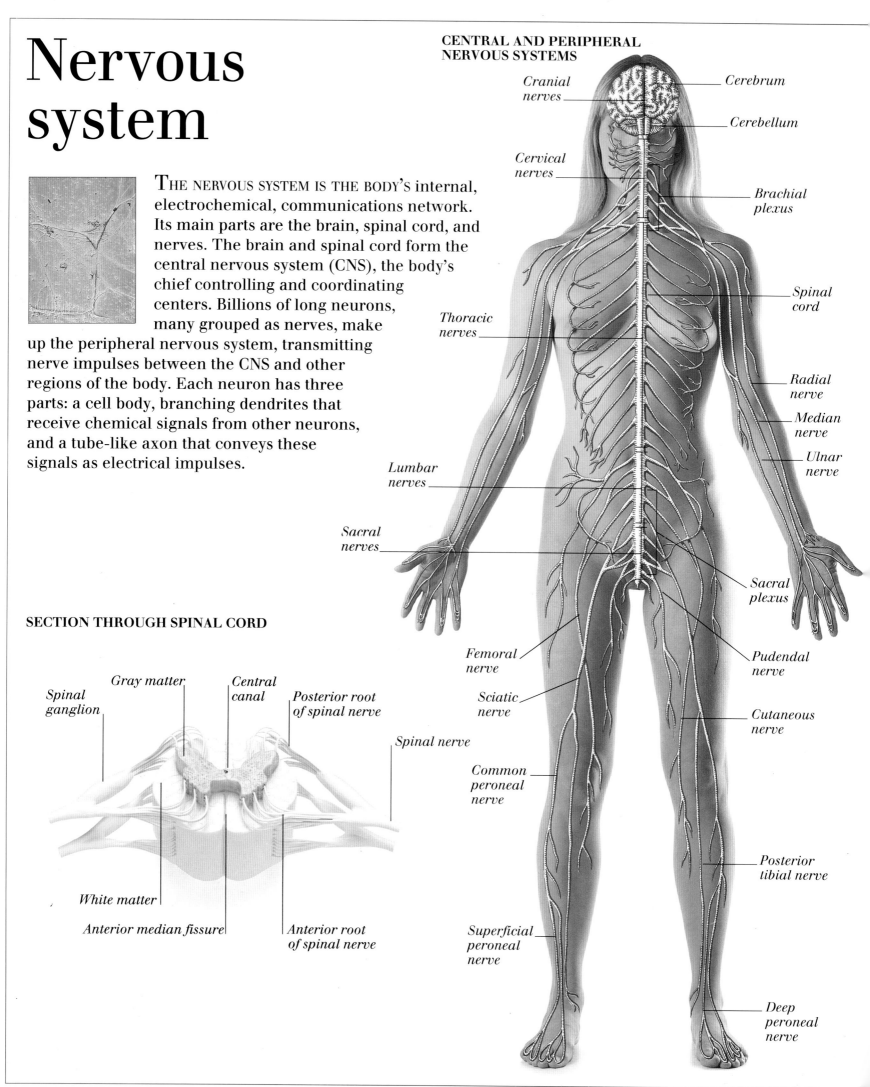

THE NERVOUS SYSTEM IS THE BODY'S internal, electrochemical, communications network. Its main parts are the brain, spinal cord, and nerves. The brain and spinal cord form the central nervous system (CNS), the body's chief controlling and coordinating centers. Billions of long neurons, many grouped as nerves, make up the peripheral nervous system, transmitting nerve impulses between the CNS and other regions of the body. Each neuron has three parts: a cell body, branching dendrites that receive chemical signals from other neurons, and a tube-like axon that conveys these signals as electrical impulses.

CENTRAL AND PERIPHERAL NERVOUS SYSTEMS

Cranial nerves

Cerebrum

Cerebellum

Cervical nerves

Brachial plexus

Thoracic nerves

Spinal cord

Radial nerve

Median nerve

Ulnar nerve

Lumbar nerves

Sacral nerves

Sacral plexus

Femoral nerve

Pudendal nerve

Sciatic nerve

Cutaneous nerve

Common peroneal nerve

Posterior tibial nerve

Superficial peroneal nerve

Deep peroneal nerve

SECTION THROUGH SPINAL CORD

Spinal ganglion

Gray matter

Central canal

Posterior root of spinal nerve

Spinal nerve

White matter

Anterior median fissure

Anterior root of spinal nerve

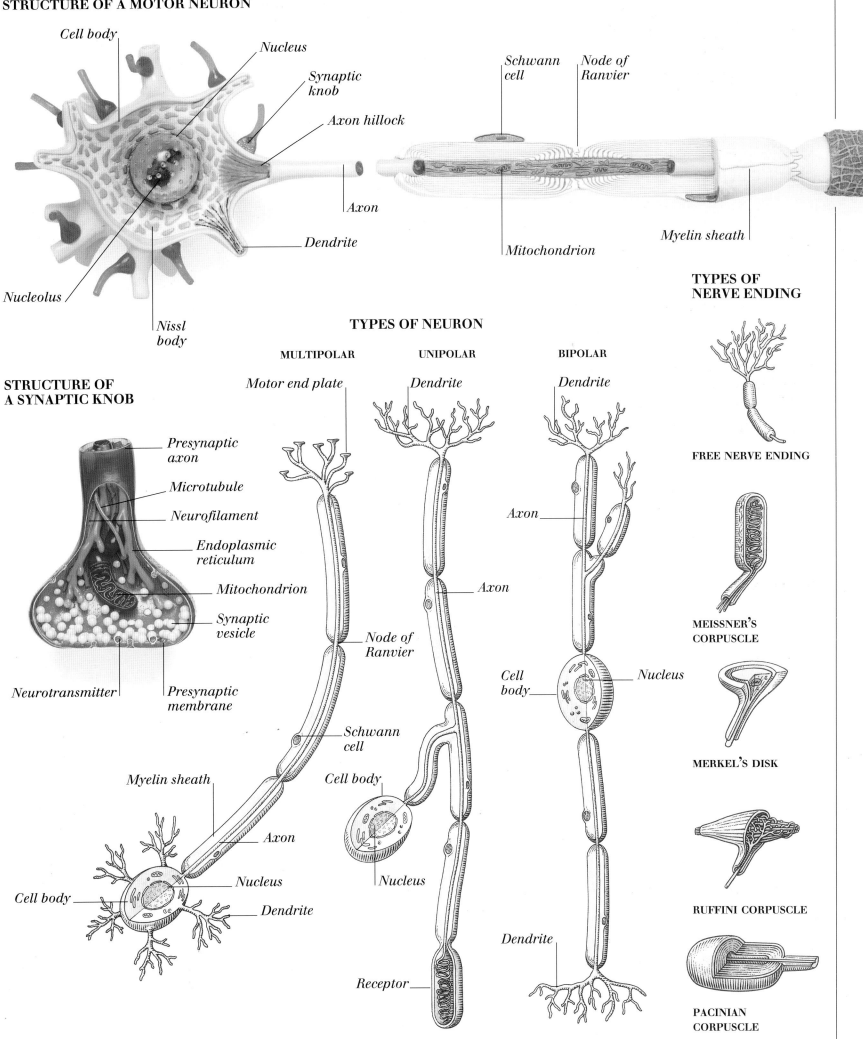

STRUCTURE OF A MOTOR NEURON

Cell body

Nucleus

Synaptic knob

Axon hillock

Axon

Dendrite

Nucleolus

Nissl body

Schwann cell

Node of Ranvier

Mitochondrion

Myelin sheath

TYPES OF NERVE ENDING

FREE NERVE ENDING

MEISSNER'S CORPUSCLE

MERKEL'S DISK

RUFFINI CORPUSCLE

PACINIAN CORPUSCLE

STRUCTURE OF A SYNAPTIC KNOB

Presynaptic axon

Microtubule

Neurofilament

Endoplasmic reticulum

Mitochondrion

Synaptic vesicle

Neurotransmitter

Presynaptic membrane

TYPES OF NEURON

MULTIPOLAR

Motor end plate

Node of Ranvier

Schwann cell

Myelin sheath

Axon

Nucleus

Dendrite

Cell body

UNIPOLAR

Dendrite

Axon

Cell body

Nucleus

Receptor

BIPOLAR

Dendrite

Axon

Cell body

Nucleus

Dendrite

Eye

THE EYE IS THE ORGAN OF SIGHT. The two eyeballs, protected within bony sockets called orbits and on the outside by the eyelids, eyebrows, and tear film, are directly connected to the brain by the optic nerves. Each eye is moved by six muscles, which are attached around the eyeball. Light rays entering the eye through the pupil are focused by the cornea and lens to form an image on the retina. The retina contains millions of light-sensitive cells, called rods and cones, which convert the image into a pattern of nerve impulses. These impulses are transmitted along the optic nerve to the brain. Information from the two optic nerves is processed in the brain to produce a single coordinated image.

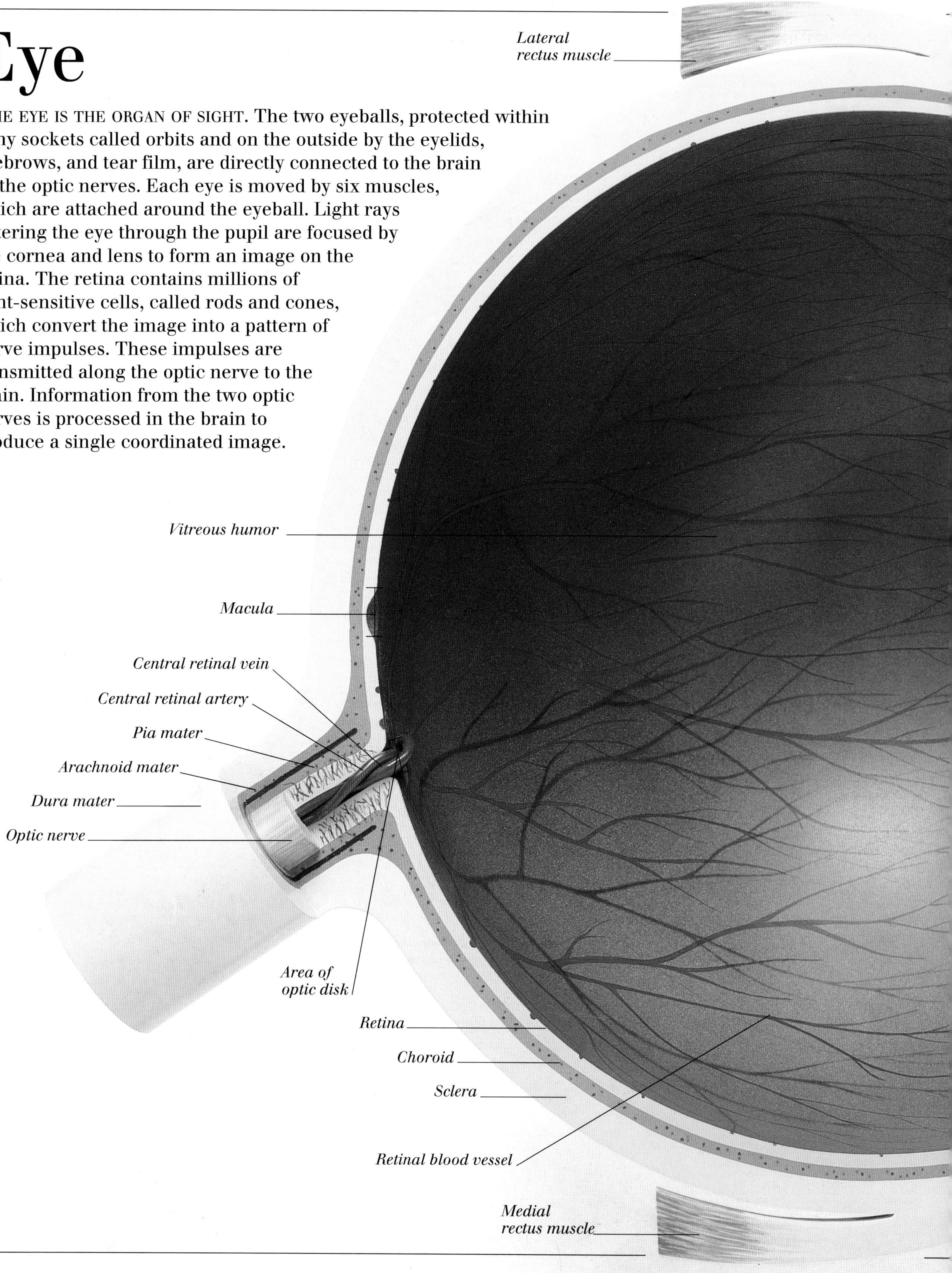

Lateral rectus muscle

Vitreous humor

Macula

Central retinal vein

Central retinal artery

Pia mater

Arachnoid mater

Dura mater

Optic nerve

Area of optic disk

Retina

Choroid

Sclera

Retinal blood vessel

Medial rectus muscle

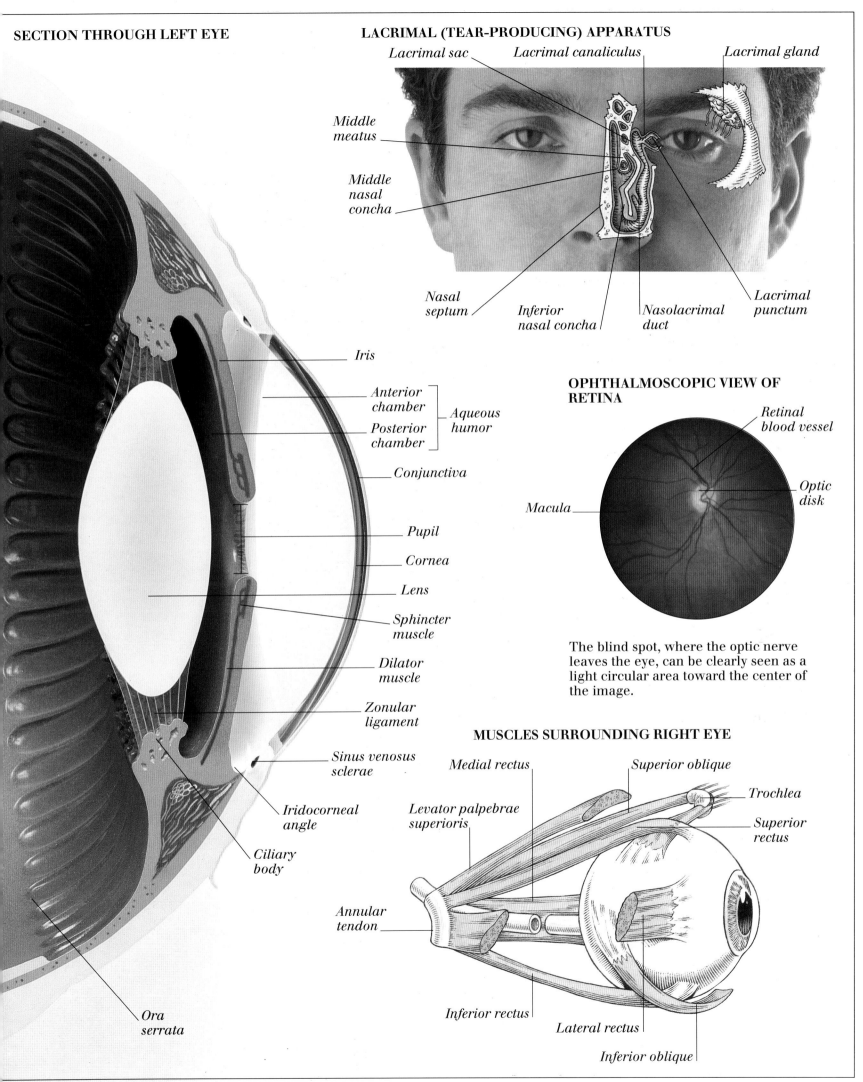

SECTION THROUGH LEFT EYE

Iris

Anterior chamber

Posterior chamber

Aqueous humor

Conjunctiva

Pupil

Cornea

Lens

Sphincter muscle

Dilator muscle

Zonular ligament

Sinus venosus sclerae

Iridocorneal angle

Ciliary body

Ora serrata

LACRIMAL (TEAR-PRODUCING) APPARATUS

Lacrimal sac

Lacrimal canaliculus

Lacrimal gland

Middle meatus

Middle nasal concha

Nasal septum

Inferior nasal concha

Nasolacrimal duct

Lacrimal punctum

OPHTHALMOSCOPIC VIEW OF RETINA

Retinal blood vessel

Optic disk

Macula

The blind spot, where the optic nerve leaves the eye, can be clearly seen as a light circular area toward the center of the image.

MUSCLES SURROUNDING RIGHT EYE

Medial rectus

Superior oblique

Trochlea

Levator palpebrae superioris

Superior rectus

Annular tendon

Inferior rectus

Lateral rectus

Inferior oblique

Ear

THE EAR IS THE ORGAN OF HEARING AND BALANCE. The outer ear
consists of a flap called the auricle or pinna and the auditory canal.
The main functional parts—the middle and inner ears—are
enclosed within the skull. The middle ear consists of three tiny
bones, known as auditory ossicles, and the eustachian tube, which
links the ear to the back of the nose. The inner ear consists of the
spiral-shaped cochlea, and also the semicircular canals and the
vestibule, which are the organs of balance. Sound waves entering
the ear travel through the auditory canal to the tympanic membrane
(eardrum), where they are converted to vibrations that are
transmitted via the ossicles to the cochlea. Here, the vibrations are
converted by millions of microscopic hairs into electrical nerve
signals to be interpreted by the brain.

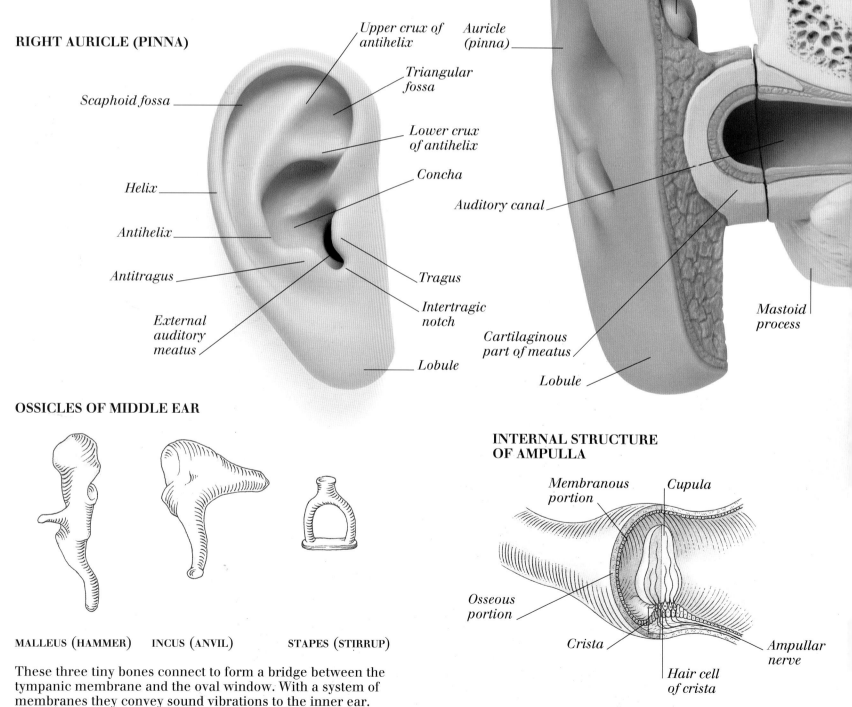

RIGHT AURICLE (PINNA)

- Upper crux of antihelix
- Auricle (pinna)
- Triangular fossa
- Scaphoid fossa
- Lower crux of antihelix
- Concha
- Helix
- Auditory canal
- Antihelix
- Antitragus
- Tragus
- Intertragic notch
- External auditory meatus
- Cartilaginous part of meatus
- Lobule
- Lobule
- Temporal bone
- Cartilage of auricle
- Mastoid process

OSSICLES OF MIDDLE EAR

MALLEUS (HAMMER)   **INCUS (ANVIL)**   **STAPES (STIRRUP)**

These three tiny bones connect to form a bridge between the
tympanic membrane and the oval window. With a system of
membranes they convey sound vibrations to the inner ear.

INTERNAL STRUCTURE OF AMPULLA

- Membranous portion
- Cupula
- Osseous portion
- Crista
- Ampullar nerve
- Hair cell of crista

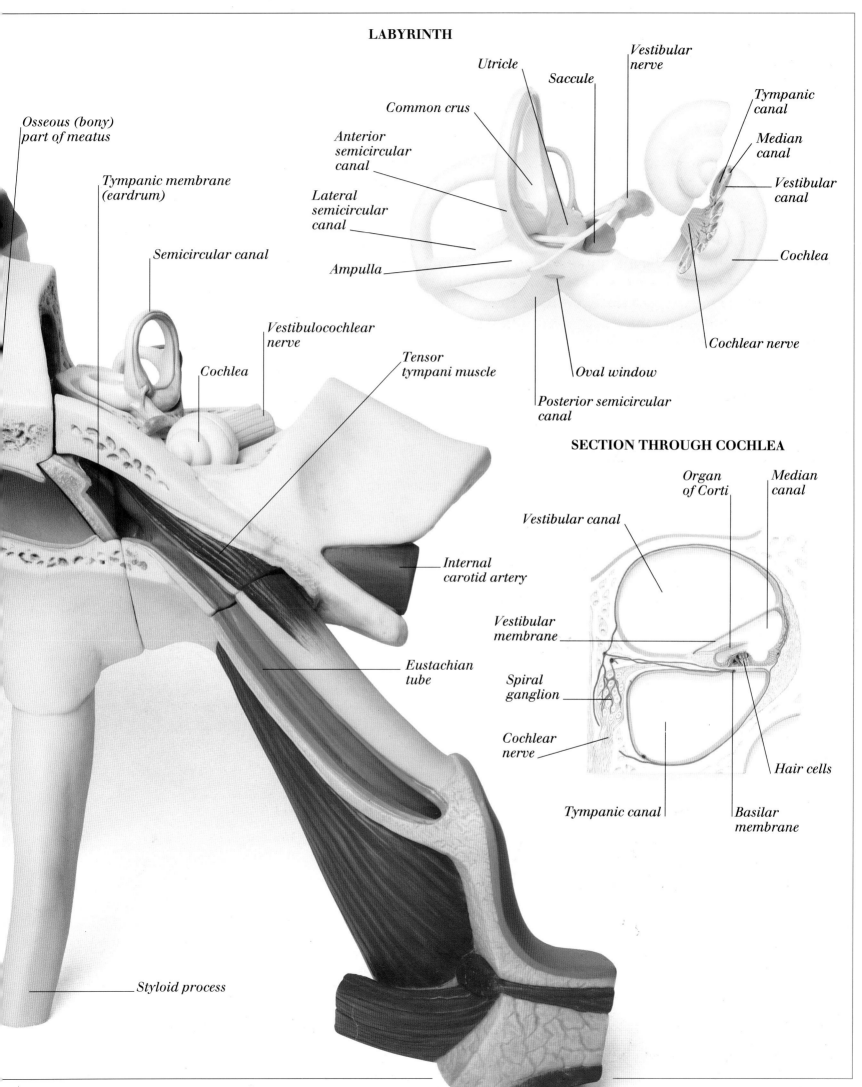

LABYRINTH

Utricle

Saccule

Vestibular nerve

Common crus

Tympanic canal

Anterior semicircular canal

Median canal

Osseous (bony) part of meatus

Vestibular canal

Lateral semicircular canal

Tympanic membrane (eardrum)

Semicircular canal

Cochlea

Ampulla

Vestibulocochlear nerve

Cochlear nerve

Cochlea

Tensor tympani muscle

Oval window

Posterior semicircular canal

SECTION THROUGH COCHLEA

Internal carotid artery

Organ of Corti

Median canal

Vestibular canal

Eustachian tube

Vestibular membrane

Spiral ganglion

Cochlear nerve

Hair cells

Tympanic canal

Basilar membrane

Styloid process

Nose, mouth, and throat

WITH EVERY BREATH, air passes through the nasal cavity down the pharynx (throat), larynx ("voice box"), and trachea (windpipe) to the lungs. The nasal cavity warms and moistens air, and the tiny layers in its lining protect the airway against damage by foreign bodies. During swallowing, the tongue moves up and back, the larynx rises, the epiglottis closes off the entrance to the trachea, and the soft palate separates the nasal cavity from the pharynx. Saliva, secreted from three pairs of salivary glands, lubricates food to make swallowing easier; it also begins the chemical breakdown of food, and helps to produce taste. The senses of taste and smell are closely linked. Both depend on the detection of dissolved molecules by sensory receptors in the olfactory nerve endings of the nose and in the taste buds of the tongue.

STRUCTURE OF TONGUE

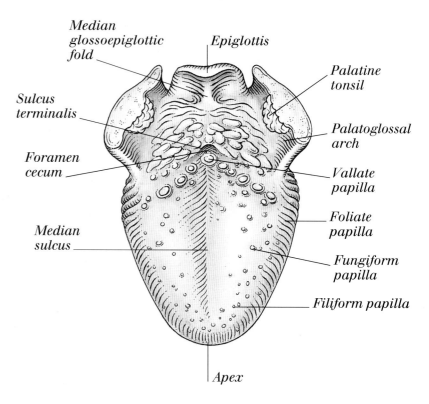

- Median glossoepiglottic fold
- Epiglottis
- Palatine tonsil
- Sulcus terminalis
- Palatoglossal arch
- Foramen cecum
- Vallate papilla
- Foliate papilla
- Median sulcus
- Fungiform papilla
- Filiform papilla
- Apex

TASTE AREAS ON TONGUE

STRUCTURES SURROUNDING PHARYNX

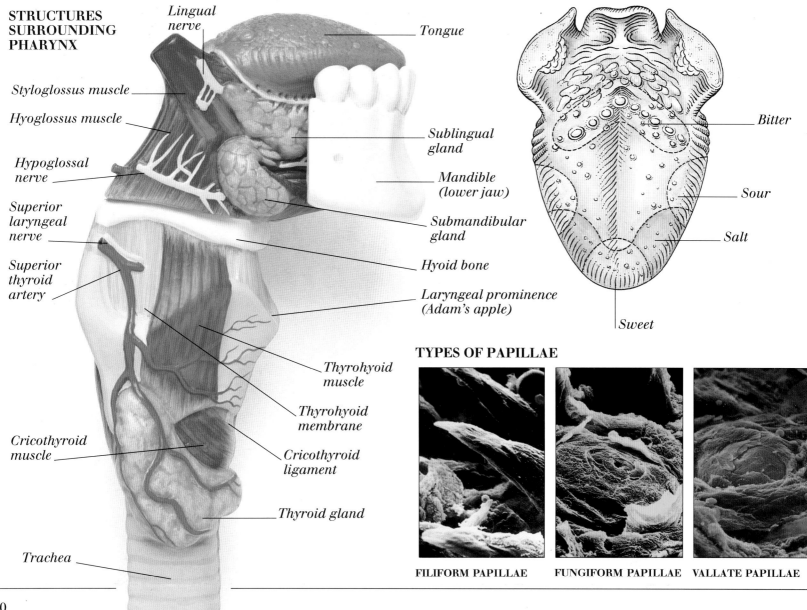

- Lingual nerve
- Tongue
- Styloglossus muscle
- Hyoglossus muscle
- Sublingual gland
- Hypoglossal nerve
- Mandible (lower jaw)
- Superior laryngeal nerve
- Submandibular gland
- Superior thyroid artery
- Hyoid bone
- Laryngeal prominence (Adam's apple)
- Thyrohyoid muscle
- Thyrohyoid membrane
- Cricothyroid muscle
- Cricothyroid ligament
- Thyroid gland
- Trachea

Taste areas: Bitter, Sour, Salt, Sweet

TYPES OF PAPILLAE

FILIFORM PAPILLAE   **FUNGIFORM PAPILLAE**   **VALLATE PAPILLAE**

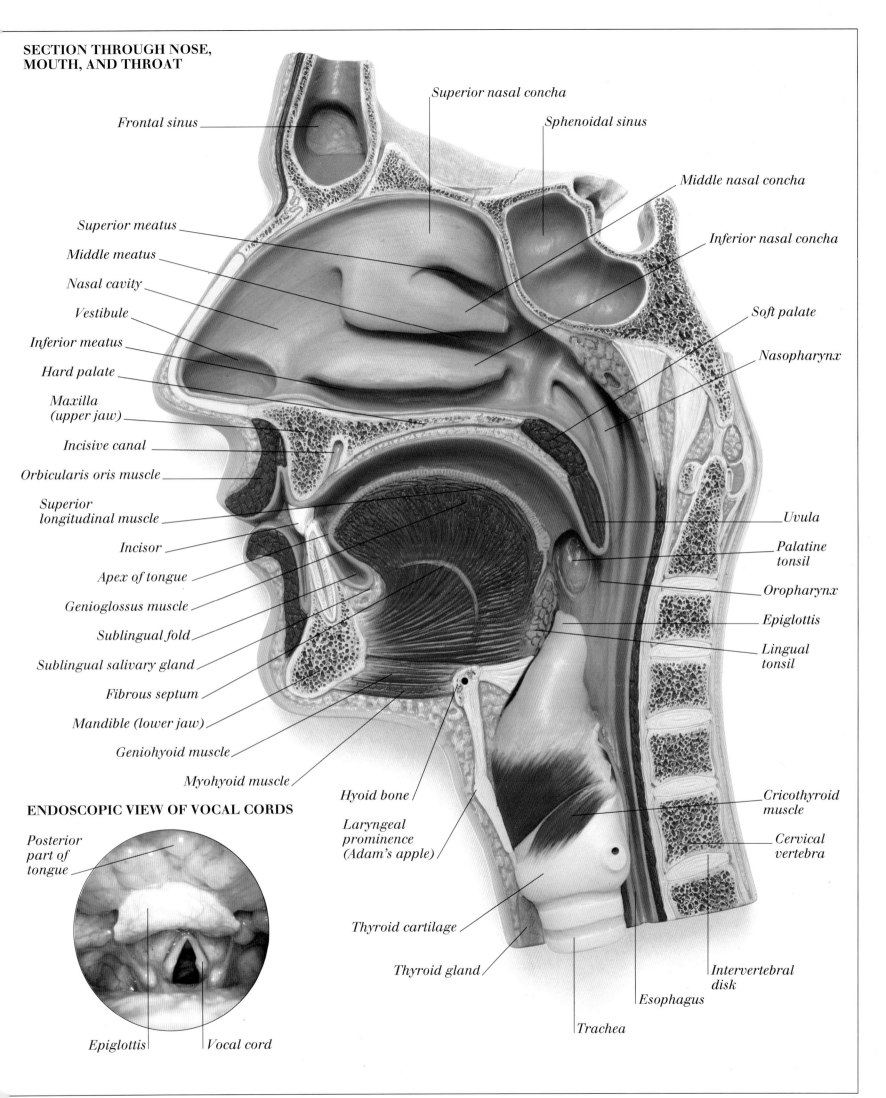

SECTION THROUGH NOSE, MOUTH, AND THROAT

Frontal sinus

Superior nasal concha

Sphenoidal sinus

Middle nasal concha

Inferior nasal concha

Superior meatus

Middle meatus

Nasal cavity

Vestibule

Inferior meatus

Hard palate

Soft palate

Nasopharynx

Maxilla (upper jaw)

Incisive canal

Orbicularis oris muscle

Superior longitudinal muscle

Uvula

Incisor

Palatine tonsil

Apex of tongue

Oropharynx

Genioglossus muscle

Epiglottis

Sublingual fold

Lingual tonsil

Sublingual salivary gland

Fibrous septum

Mandible (lower jaw)

Geniohyoid muscle

Myohyoid muscle

Hyoid bone

Cricothyroid muscle

Cervical vertebra

Laryngeal prominence (Adam's apple)

ENDOSCOPIC VIEW OF VOCAL CORDS

Posterior part of tongue

Thyroid cartilage

Intervertebral disk

Thyroid gland

Esophagus

Epiglottis

Vocal cord

Trachea

Teeth

THE 20 PRIMARY TEETH (also called deciduous or milk
teeth) usually begin to erupt when a baby is about six
months old. They start to be replaced by the permanent
teeth when the child is about six years old. By the age of
20, most adults have a full set of 32 teeth although the
third molars (commonly called wisdom teeth) may never
erupt. While teeth help people to speak clearly and give
shape to the face, their main function is the chewing of
food. Incisors and canines shear and tear the food into
pieces; premolars and molars crush and grind it further.
Although tooth enamel is the hardest substance in the
body, it tends to be eroded and destroyed by acid
produced in the mouth during the breakdown of food.

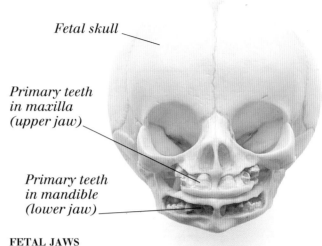

Fetal skull

*Primary teeth
in maxilla
(upper jaw)*

*Primary teeth
in mandible
(lower jaw)*

FETAL JAWS
By the sixth week of embryonic development areas of
thickening occur in each jaw; these areas give rise to
tooth buds. By the time the fetus is six months old,
enamel has formed on the tooth buds.

DEVELOPMENT OF JAW AND TEETH

*Maxilla
(upper jaw)*

*Mandible
(lower jaw)*

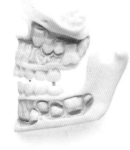
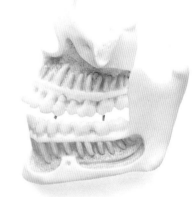

A NEWBORN BABY'S JAWS
The primary teeth can be seen
developing in the jawbones;
they begin to erupt around the
age of six months.

A FIVE-YEAR-OLD CHILD'S TEETH
There is a full set of 20 erupted
primary teeth; the permanent
teeth can be seen developing in
the upper and lower jaws.

A NINE-YEAR-OLD CHILD'S TEETH
Most of the teeth are primary
teeth but the permanent
incisors and first molars have
now emerged.

AN ADULT'S TEETH
By the age of 20, the full set of
32 permanent teeth (including
the wisdom teeth) should be
in position.

THE PERMANENT TEETH

| *Molars* | | | *Premolars* | | *Canines* | *Incisors* | | | *Canines* | *Premolars* | | *Molars* | | |

UPPER

LOWER

| 3rd (wisdom) | 2nd | 1st | 2nd | 1st | | Lateral | Central | Lateral | | 1st | 2nd | 1st | 2nd | 3rd (wisdom) |

STRUCTURE OF A TOOTH

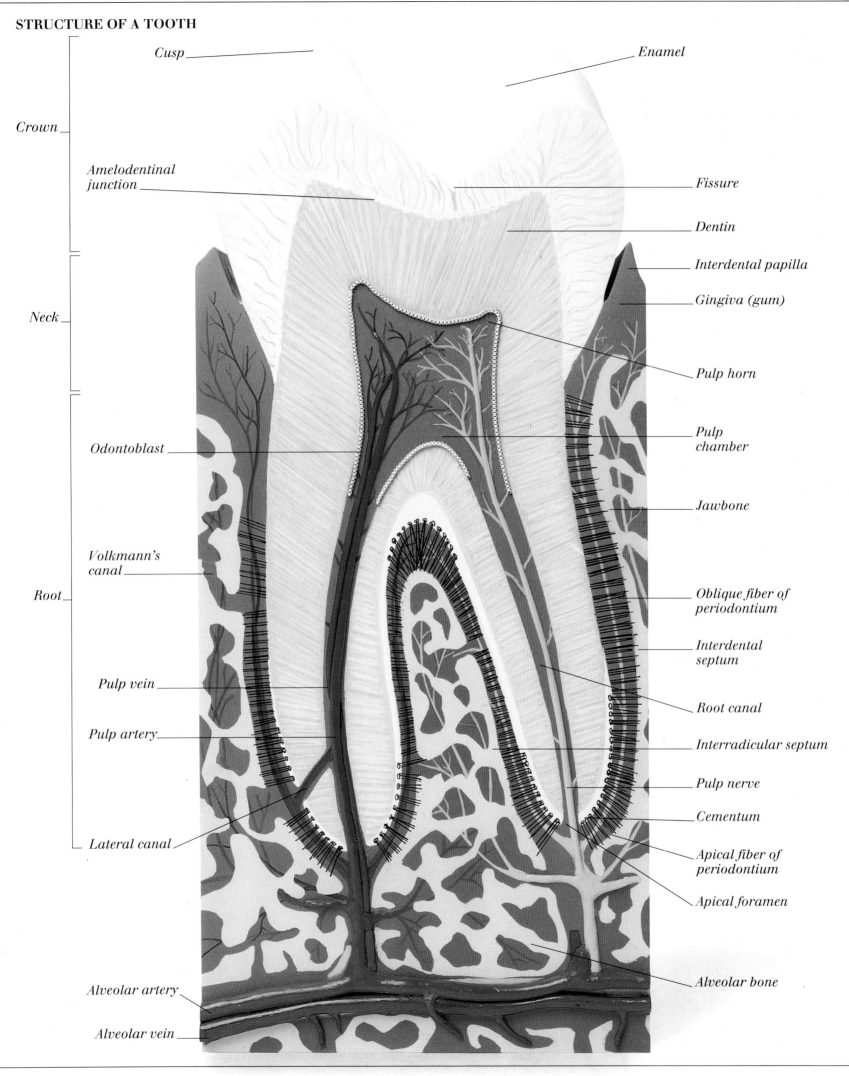

Cusp

Enamel

Crown

Amelodentinal
junction

Fissure

Dentin

Neck

Interdental papilla

Gingiva (gum)

Pulp horn

Odontoblast

Pulp
chamber

Jawbone

Volkmann's
canal

Oblique fiber of
periodontium

Root

Interdental
septum

Pulp vein

Root canal

Pulp artery

Interradicular septum

Pulp nerve

Cementum

Lateral canal

Apical fiber of
periodontium

Apical foramen

Alveolar artery

Alveolar bone

Alveolar vein

Digestive system 1

THE DIGESTIVE SYSTEM BREAKS DOWN FOOD into particles so tiny that blood can take nourishment to all parts of the body. The system's main part is a 30-foot (9 m) tube from mouth to rectum; muscles in this alimentary canal force food along. Chewed food first travels through the esophagus to the stomach, which churns and liquidizes food before it passes through the duodenum, jejunum, and ileum—the three parts of the long, convoluted small intestine. Here, digestive juices from the gallbladder and pancreas break down food particles; many filter out into the blood through tiny fingerlike villi that line the small intestine's inner wall. Undigested food in the colon forms feces that leave the body through the anus.

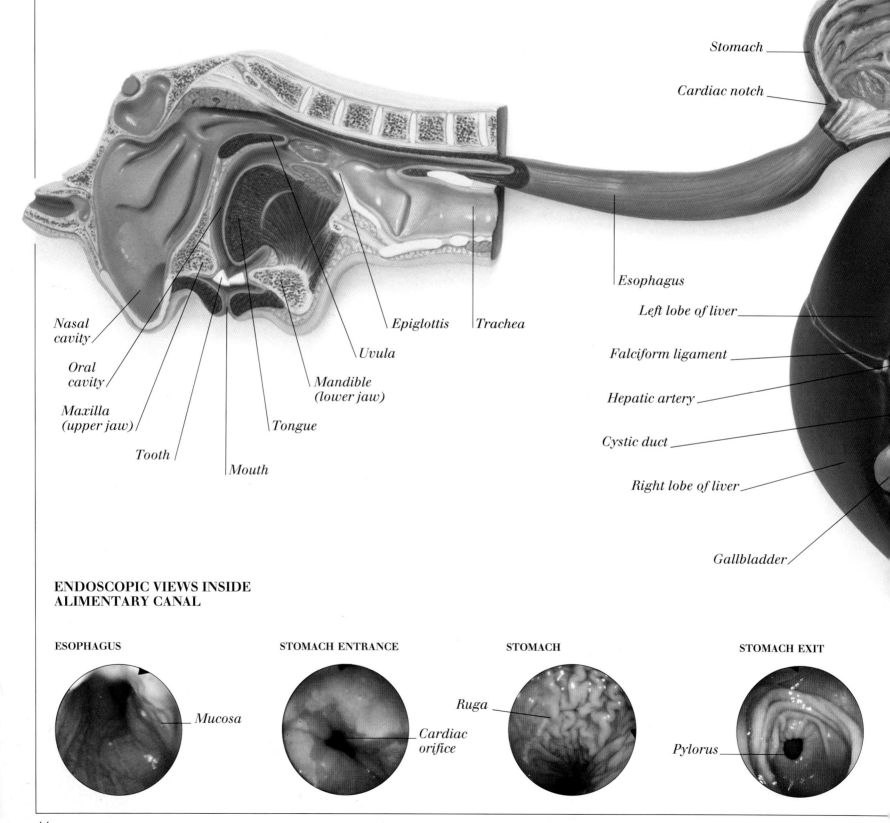

Stomach

Cardiac notch

Nasal cavity

Oral cavity

Maxilla (upper jaw)

Tooth

Mouth

Tongue

Mandible (lower jaw)

Uvula

Epiglottis

Trachea

Esophagus

Left lobe of liver

Falciform ligament

Hepatic artery

Cystic duct

Right lobe of liver

Gallbladder

ENDOSCOPIC VIEWS INSIDE ALIMENTARY CANAL

ESOPHAGUS

Mucosa

STOMACH ENTRANCE

Cardiac orifice

STOMACH

Ruga

STOMACH EXIT

Pylorus

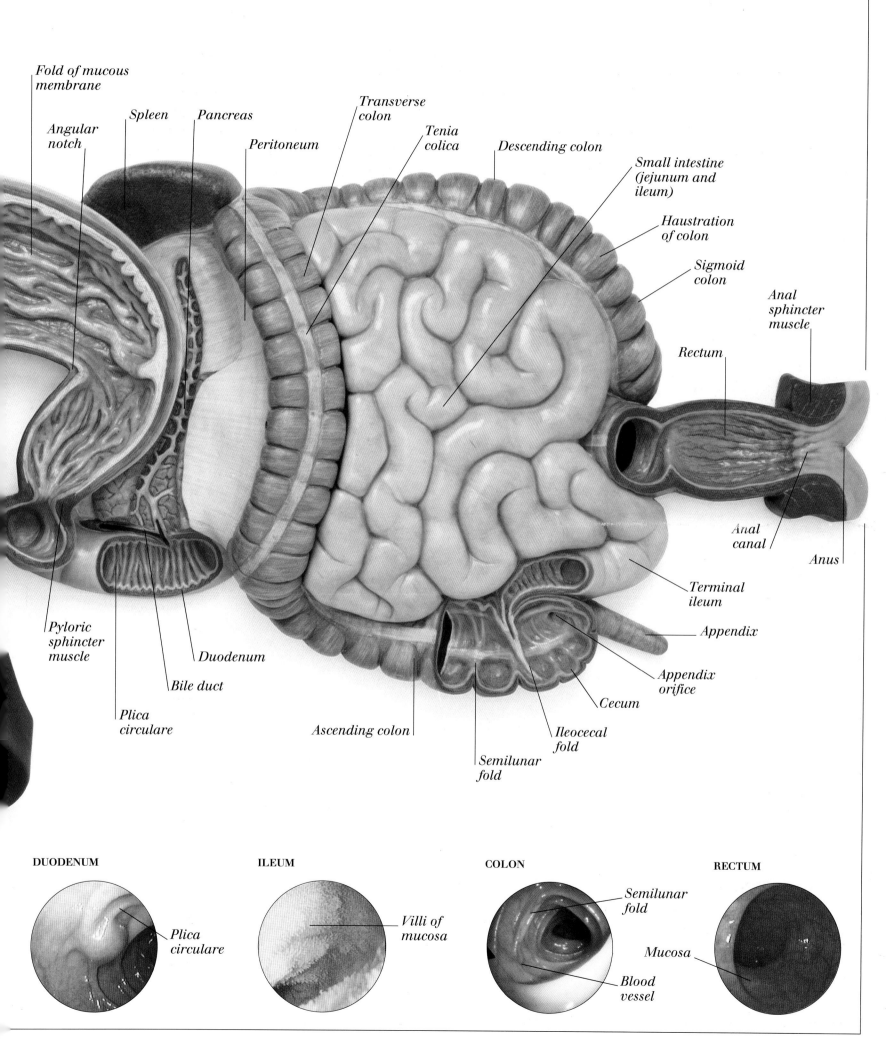

Fold of mucous membrane

Angular notch

Spleen

Pancreas

Peritoneum

Transverse colon

Tenia colica

Descending colon

Small intestine (jejunum and ileum)

Haustration of colon

Sigmoid colon

Anal sphincter muscle

Rectum

Pyloric sphincter muscle

Bile duct

Duodenum

Plica circulare

Ascending colon

Semilunar fold

Ileocecal fold

Cecum

Appendix orifice

Appendix

Terminal ileum

Anal canal

Anus

DUODENUM

Plica circulare

ILEUM

Villi of mucosa

COLON

Semilunar fold

Blood vessel

RECTUM

Mucosa

Digestive system 2

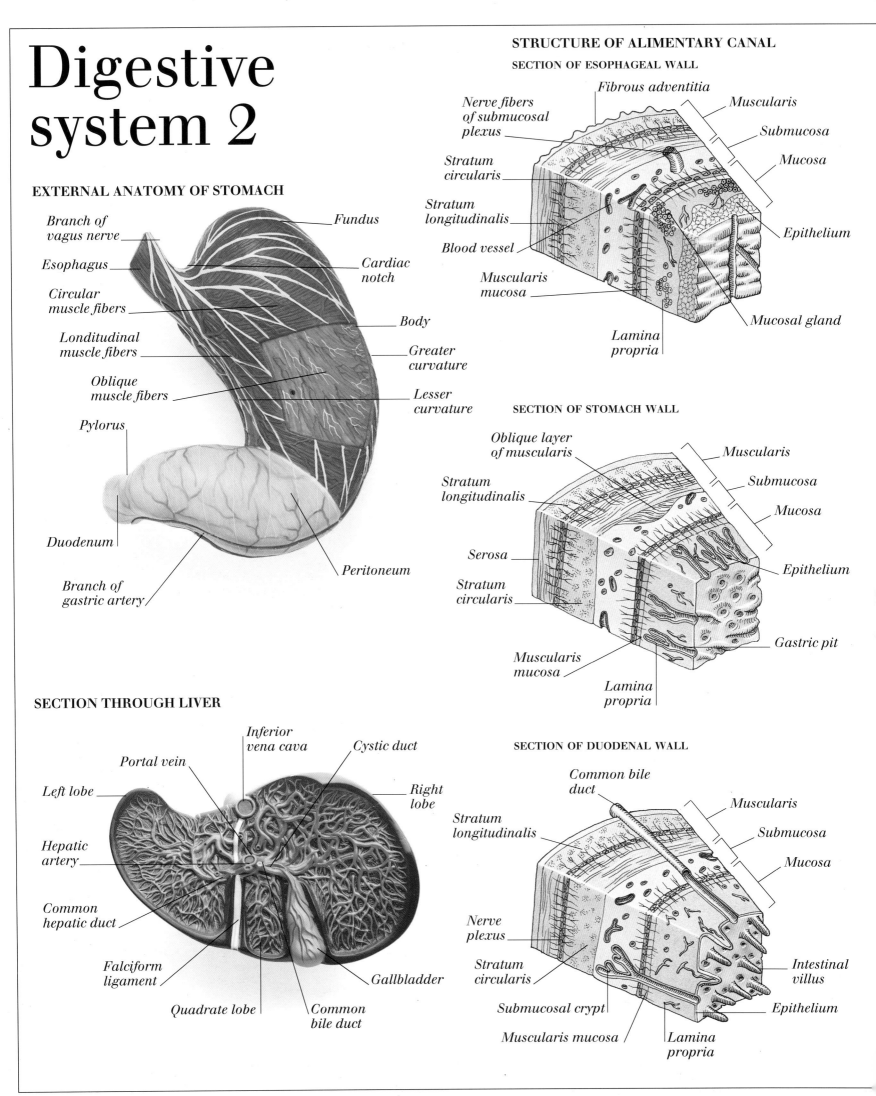

EXTERNAL ANATOMY OF STOMACH

Branch of vagus nerve

Esophagus

Circular muscle fibers

Londitudinal muscle fibers

Oblique muscle fibers

Pylorus

Duodenum

Branch of gastric artery

Fundus

Cardiac notch

Body

Greater curvature

Lesser curvature

Peritoneum

STRUCTURE OF ALIMENTARY CANAL

SECTION OF ESOPHAGEAL WALL

Nerve fibers of submucosal plexus

Stratum circularis

Stratum longitudinalis

Blood vessel

Muscularis mucosa

Lamina propria

Fibrous adventitia

Muscularis

Submucosa

Mucosa

Epithelium

Mucosal gland

SECTION OF STOMACH WALL

Oblique layer of muscularis

Stratum longitudinalis

Serosa

Stratum circularis

Muscularis mucosa

Lamina propria

Muscularis

Submucosa

Mucosa

Epithelium

Gastric pit

SECTION THROUGH LIVER

Portal vein

Left lobe

Hepatic artery

Common hepatic duct

Falciform ligament

Quadrate lobe

Inferior vena cava

Cystic duct

Right lobe

Common bile duct

Gallbladder

SECTION OF DUODENAL WALL

Common bile duct

Stratum longitudinalis

Nerve plexus

Stratum circularis

Submucosal crypt

Muscularis mucosa

Muscularis

Submucosa

Mucosa

Intestinal villus

Epithelium

Lamina propria

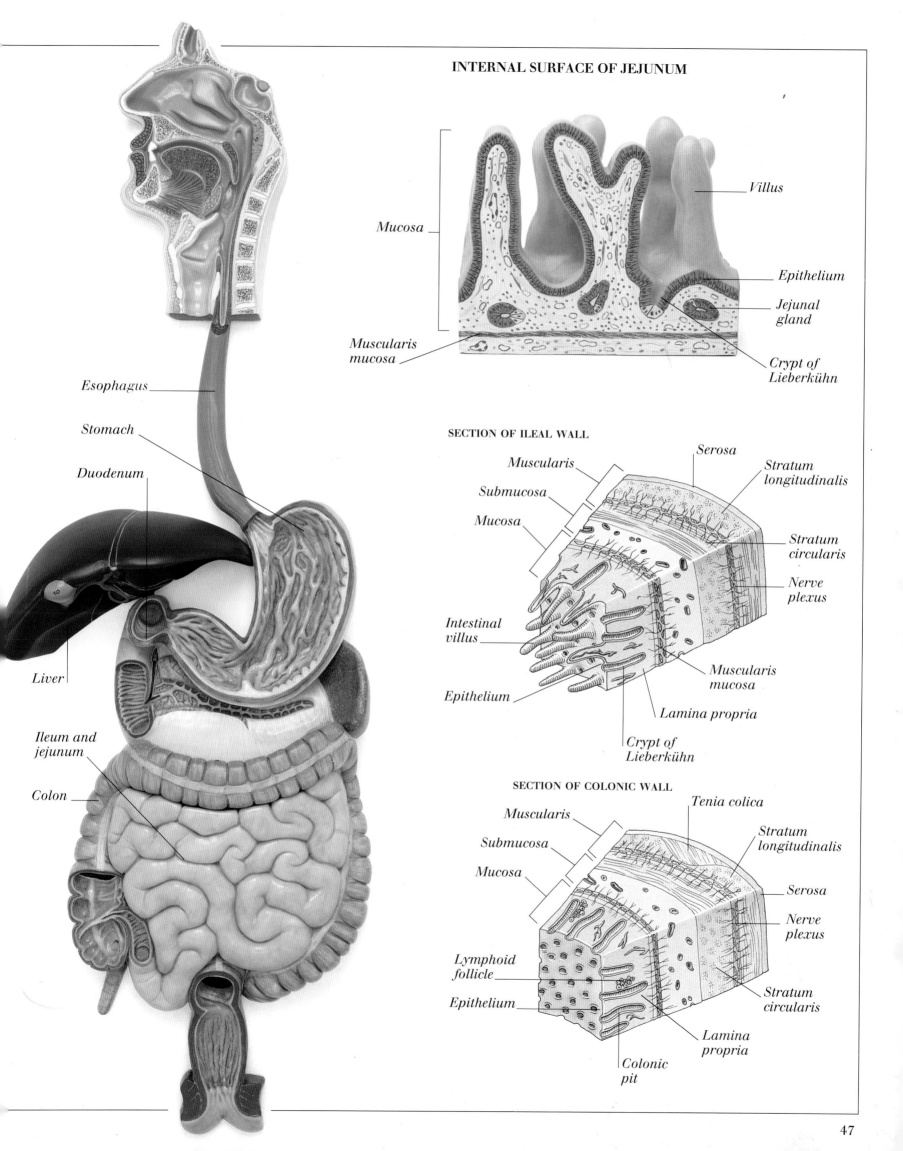

INTERNAL SURFACE OF JEJUNUM

Villus

Mucosa

Epithelium

Jejunal gland

Muscularis mucosa

Crypt of Lieberkühn

Esophagus

Stomach

Duodenum

SECTION OF ILEAL WALL

Serosa

Muscularis

Stratum longitudinalis

Submucosa

Mucosa

Stratum circularis

Nerve plexus

Intestinal villus

Liver

Muscularis mucosa

Epithelium

Lamina propria

Ileum and jejunum

Crypt of Lieberkühn

Colon

SECTION OF COLONIC WALL

Tenia colica

Muscularis

Stratum longitudinalis

Submucosa

Mucosa

Serosa

Nerve plexus

Lymphoid follicle

Epithelium

Stratum circularis

Lamina propria

Colonic pit

47

Heart

THE HEART IS A HOLLOW MUSCLE in the middle of the chest that pumps blood around the body, supplying cells with oxygen and nutrients. A muscular wall, called the septum, divides the heart lengthwise into left and right sides. A valve divides each side into two chambers: an upper atrium and a lower ventricle. When the heart muscle contracts, it squeezes blood through the atria and then through the ventricles. Oxygenated blood from the lungs flows from the pulmonary veins into the left atrium, through the left ventricle, and then out via the aorta to all parts of the body. Deoxygenated blood returning from the body flows from the vena cava into the right atrium, through the right ventricle, and then out via the pulmonary artery to the lungs for reoxygenation. At rest the heart beats between 60 and 80 times a minute; during exercise or at times of stress or excitement the rate may increase to 200 beats a minute.

ARTERIES AND VEINS SURROUNDING HEART

Aorta

Left coronary artery

Cardiac vein

Right coronary artery

Coronary sinus

Main branch of left coronary artery

SECTION THROUGH HEART WALL

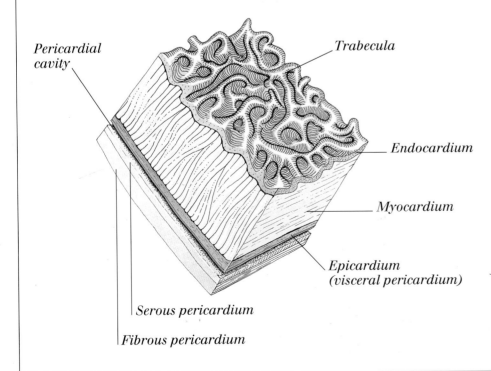

Pericardial cavity

Trabecula

Endocardium

Myocardium

Epicardium (visceral pericardium)

Serous pericardium

Fibrous pericardium

HEARTBEAT SEQUENCE

ATRIAL DIASTOLE

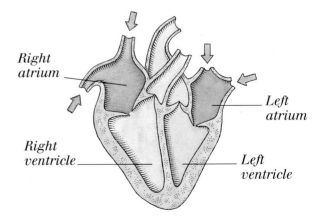

Right atrium

Left atrium

Right ventricle

Left ventricle

Deoxygenated blood enters the right atrium while the left atrium receives oxygenated blood.

STRUCTURE OF HEART

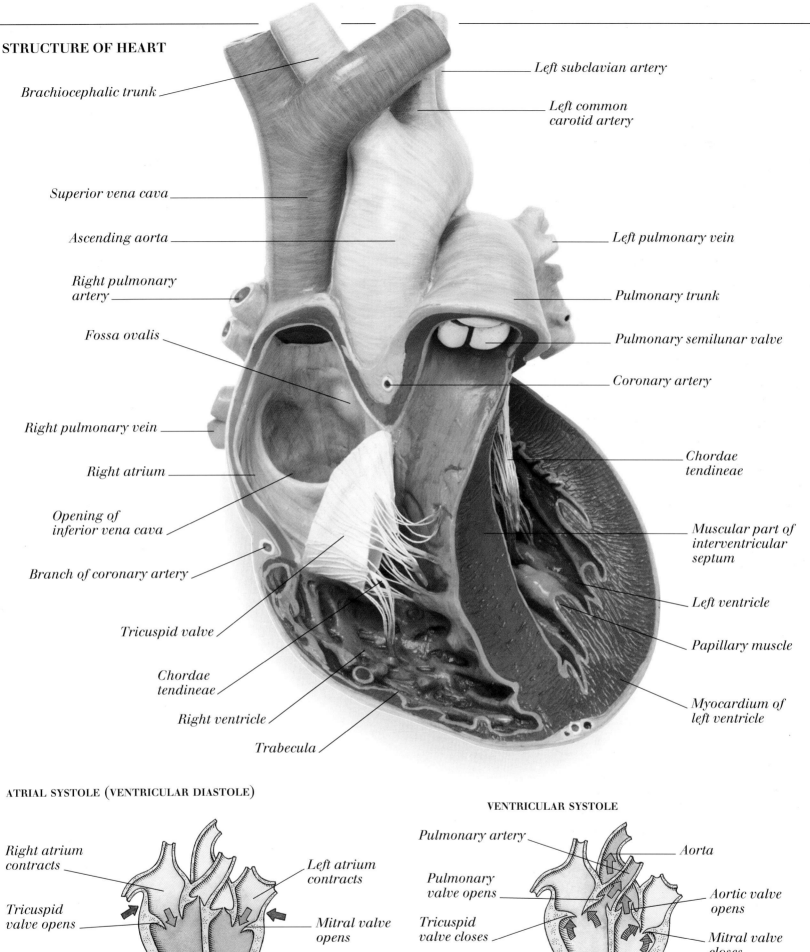

Brachiocephalic trunk

Superior vena cava

Ascending aorta

Right pulmonary artery

Fossa ovalis

Right pulmonary vein

Right atrium

Opening of inferior vena cava

Branch of coronary artery

Tricuspid valve

Chordae tendineae

Right ventricle

Trabecula

Left subclavian artery

Left common carotid artery

Left pulmonary vein

Pulmonary trunk

Pulmonary semilunar valve

Coronary artery

Chordae tendineae

Muscular part of interventricular septum

Left ventricle

Papillary muscle

Myocardium of left ventricle

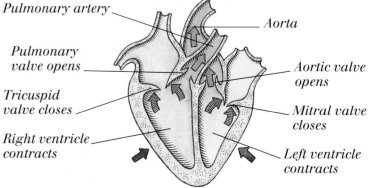

ATRIAL SYSTOLE (VENTRICULAR DIASTOLE)

Right atrium contracts

Tricuspid valve opens

Right ventricle dilates

Left atrium contracts

Mitral valve opens

Left ventricle dilates

Left and right atria contract, forcing blood into the relaxed ventricles.

VENTRICULAR SYSTOLE

Pulmonary artery

Pulmonary valve opens

Tricuspid valve closes

Right ventricle contracts

Aorta

Aortic valve opens

Mitral valve closes

Left ventricle contracts

Ventricles contract and force blood to the lungs for oxygenation and via the aorta to the rest of the body.

Circulatory system

THE CIRCULATORY SYSTEM consists of the heart and blood vessels, which together maintain a continuous flow of blood around the body. The heart pumps oxygen-rich blood from the lungs to all parts of the body through a network of tubes called arteries, and smaller branches called arterioles. Blood returns to the heart via small vessels called venules, which lead in turn into larger tubes called veins. Arterioles and venules are linked by a network of tiny vessels called capillaries, where the exchange of oxygen and carbon dioxide between blood and body cells takes place. Blood has four main components: red blood cells, white blood cells, platelets, and liquid plasma.

ARTERIAL SYSTEM OF BRAIN

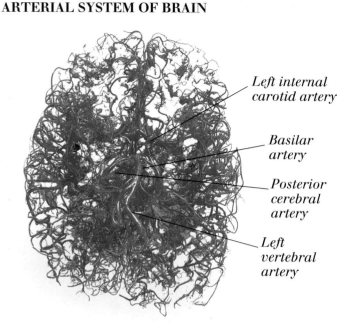

Left internal carotid artery

Basilar artery

Posterior cerebral artery

Left vertebral artery

CIRCULATORY SYSTEM OF LIVER

Inferior vena cava

Portal vein

Common bile duct

Hepatic artery

Gallbladder

CIRCULATORY SYSTEM OF HEART AND LUNGS

Superior vena cava

Aorta

Right ventricle

Left ventricle

SECTION OF MAIN ARTERY

Tunica media

Collagen and elastic fibers

External elastic lamina

Tunica adventitia

Internal elastic lamina

Tunica intima

Endothelium

Arteriole

SECTION OF MAIN VEIN

Tunica media

Collagen and elastic fibers

External elastic lamina

Tunica adventitia

Valve cusp

Internal elastic lamina

Tunica intima

Endothelium

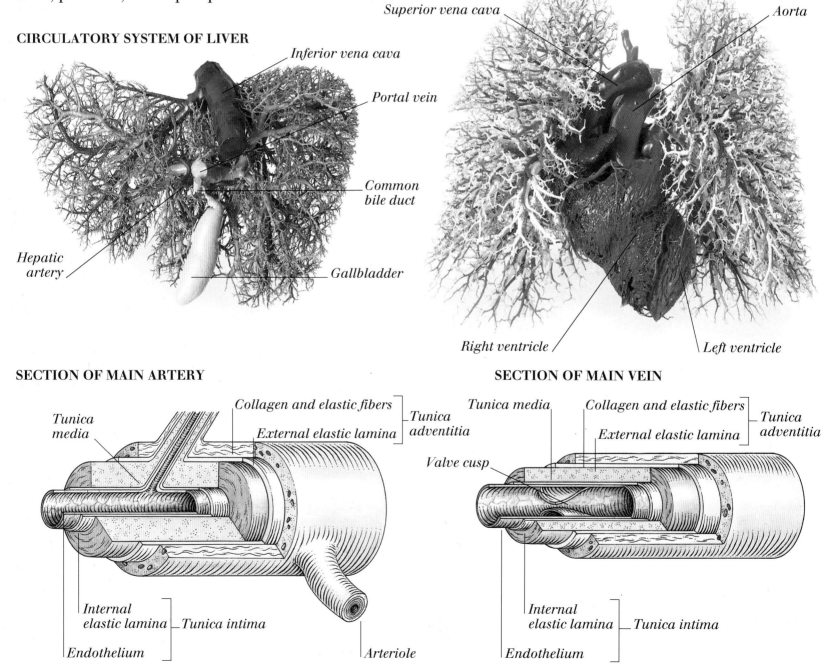

PRINCIPAL ARTERIES AND VEINS OF CIRCULATORY SYSTEM

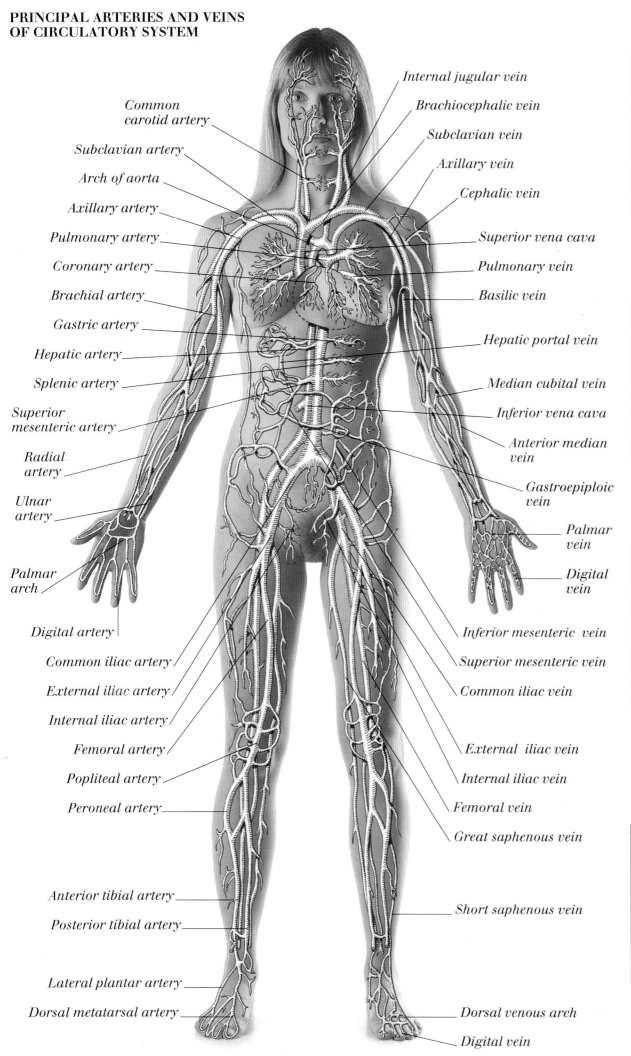

Common carotid artery

Subclavian artery

Arch of aorta

Axillary artery

Pulmonary artery

Coronary artery

Brachial artery

Gastric artery

Hepatic artery

Splenic artery

Superior mesenteric artery

Radial artery

Ulnar artery

Palmar arch

Digital artery

Common iliac artery

External iliac artery

Internal iliac artery

Femoral artery

Popliteal artery

Peroneal artery

Anterior tibial artery

Posterior tibial artery

Lateral plantar artery

Dorsal metatarsal artery

Internal jugular vein

Brachiocephalic vein

Subclavian vein

Axillary vein

Cephalic vein

Superior vena cava

Pulmonary vein

Basilic vein

Hepatic portal vein

Median cubital vein

Inferior vena cava

Anterior median vein

Gastroepiploic vein

Palmar vein

Digital vein

Inferior mesenteric vein

Superior mesenteric vein

Common iliac vein

External iliac vein

Internal iliac vein

Femoral vein

Great saphenous vein

Short saphenous vein

Dorsal venous arch

Digital vein

TYPES OF BLOOD CELLS

RED BLOOD CELLS
These cells are biconcave in shape to maximize their oxygen-carrying capacity.

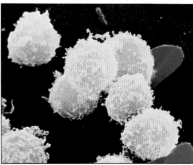

WHITE BLOOD CELLS
Lymphocytes are the smallest white blood cells; they form antibodies against disease.

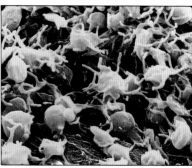

PLATELETS
Tiny cells that are activated whenever blood clotting or repair to vessels is necessary.

BLOOD CLOTTING

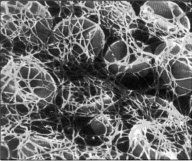

Filaments of fibrin enmesh red blood cells as part of the process of blood clotting.

Respiratory system

THE RESPIRATORY SYSTEM supplies the oxygen needed by body cells and carries off their carbon dioxide waste. Inhaled air passes via the trachea (windpipe) through two narrower tubes, the bronchi, to the lungs. Each lung comprises many fine, branching tubes called bronchioles that end in tiny clustered chambers called alveoli. Gases cross the thin alveolar walls to and from a network of tiny blood vessels. Intercostal (rib) muscles and the muscular diaphragm below the lungs operate the lungs like bellows, drawing air in and forcing it out at regular intervals.

BRONCHIOLE AND ALVEOLI

Bronchial nerve

Visceral cartilage

Branch of pulmonary vein

Terminal bronchiole

Mucosal gland

Bronchial vein

Branch of pulmonary artery

Elastic fibers

Interalveolar septum

Alveolus

Connective tissue

Capillary network

Epithelium

SEGMENTS OF BRONCHIAL TREE

Apical

Upper lobe of right lung

Posterior

Anterior

Middle lobe of right lung

Lateral

Medial

Anterior basal

Lateral basal

Lower lobe of right lung

Medial basal

Apical

Posterior basal

Apical

Posterior

Anterior

Superior lingular

Inferior lingular

Upper lobe of left lung

Apical

Medial basal

Anterior basal

Lower lobe of left lung

Lateral basal

Posterior basal

STRUCTURES OF THORACIC CAVITY

GASEOUS EXCHANGE IN ALVEOLUS

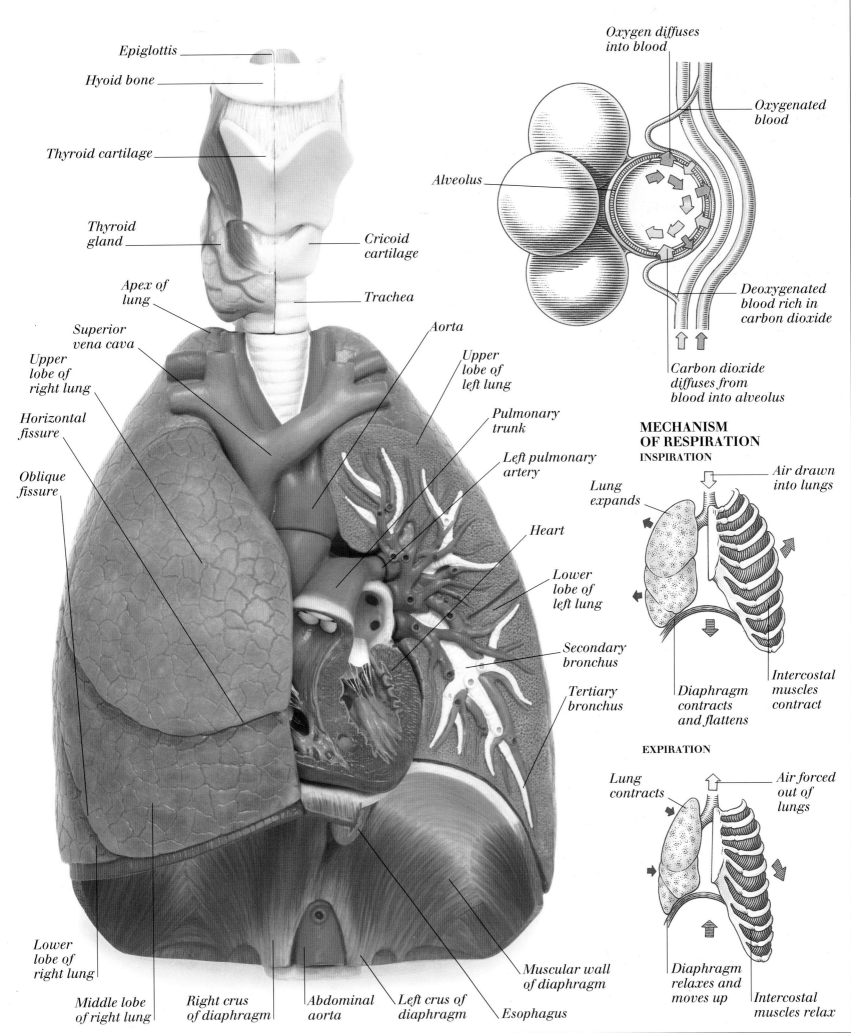

Epiglottis

Hyoid bone

Thyroid cartilage

Thyroid gland

Cricoid cartilage

Apex of lung

Trachea

Superior vena cava

Aorta

Upper lobe of right lung

Upper lobe of left lung

Horizontal fissure

Pulmonary trunk

Oblique fissure

Left pulmonary artery

Heart

Lower lobe of left lung

Secondary bronchus

Tertiary bronchus

Lower lobe of right lung

Muscular wall of diaphragm

Middle lobe of right lung

Right crus of diaphragm

Abdominal aorta

Left crus of diaphragm

Esophagus

Oxygen diffuses into blood

Oxygenated blood

Alveolus

Deoxygenated blood rich in carbon dioxide

Carbon dioxide diffuses from blood into alveolus

MECHANISM OF RESPIRATION
INSPIRATION

Lung expands

Air drawn into lungs

Diaphragm contracts and flattens

Intercostal muscles contract

EXPIRATION

Lung contracts

Air forced out of lungs

Diaphragm relaxes and moves up

Intercostal muscles relax

Urinary system

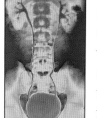

THE URINARY SYSTEM FILTERS WASTE PRODUCTS from the blood and removes them from the body via a system of tubes. Blood is filtered in the two kidneys, which are fist-sized, bean-shaped organs. The renal arteries carry blood to the kidneys; the renal veins remove blood after filtering. Each kidney contains about one million tiny units called nephrons. Each nephron is made up of a tubule and a filtering unit called a glomerulus, which consists of a collection of tiny blood vessels surrounded by the hollow Bowman's capsule. The filtering process produces a watery fluid that leaves the kidney as urine. The urine is carried via two tubes called ureters to the bladder, where it is stored until its release from the body through another tube called the urethra.

ARTERIAL SYSTEM OF KIDNEYS

Aorta

Celiac trunk

Superior mesenteric artery

Right renal artery

Left renal artery

Right ureter

Left ureter

SECTION THROUGH LEFT KIDNEY

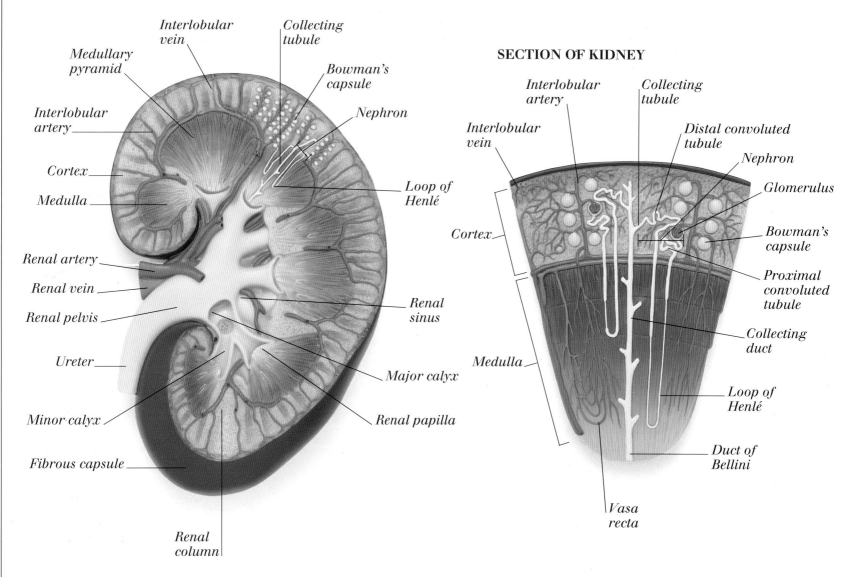

Interlobular vein

Medullary pyramid

Collecting tubule

Bowman's capsule

Nephron

Interlobular artery

Cortex

Medulla

Loop of Henlé

Renal artery

Renal vein

Renal pelvis

Renal sinus

Ureter

Major calyx

Minor calyx

Renal papilla

Fibrous capsule

Renal column

SECTION OF KIDNEY

Interlobular artery

Collecting tubule

Interlobular vein

Distal convoluted tubule

Nephron

Glomerulus

Cortex

Bowman's capsule

Proximal convoluted tubule

Medulla

Collecting duct

Loop of Henlé

Duct of Bellini

Vasa recta

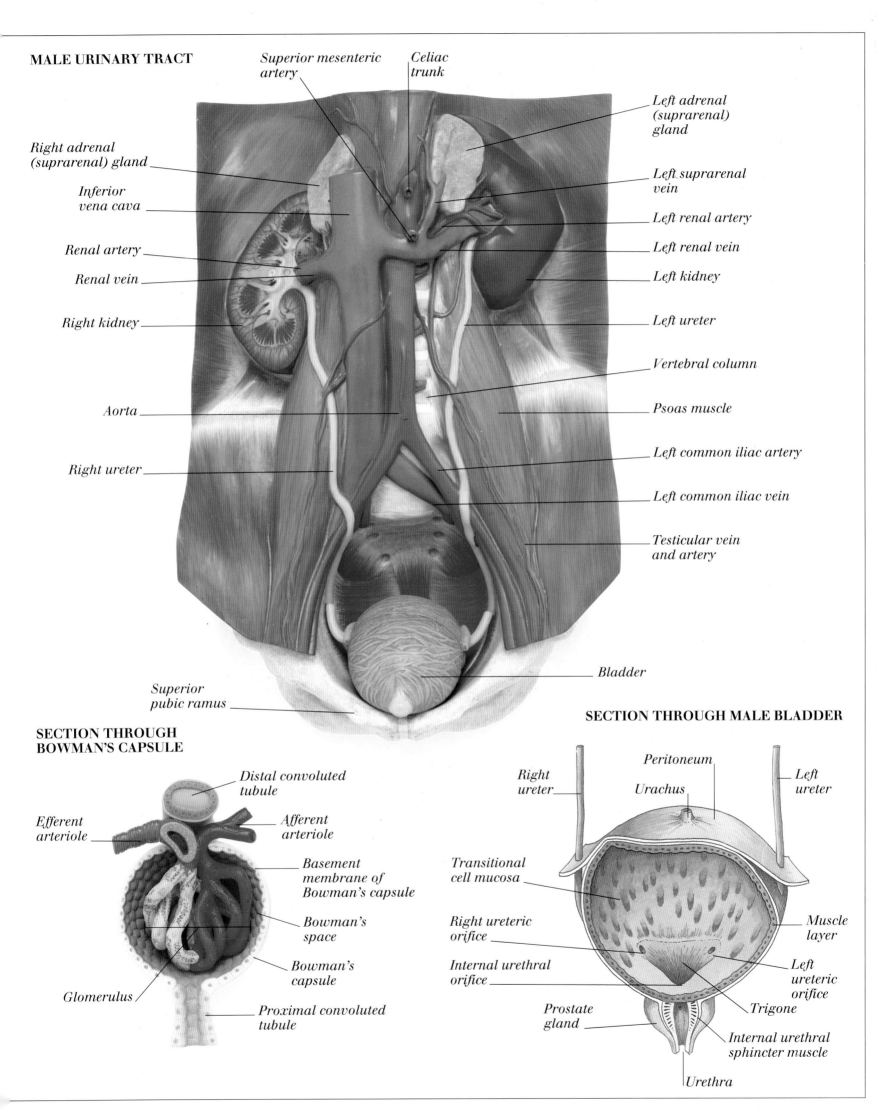

MALE URINARY TRACT

Superior mesenteric artery

Celiac trunk

Left adrenal (suprarenal) gland

Right adrenal (suprarenal) gland

Inferior vena cava

Left suprarenal vein

Left renal artery

Renal artery

Left renal vein

Renal vein

Left kidney

Right kidney

Left ureter

Vertebral column

Aorta

Psoas muscle

Right ureter

Left common iliac artery

Left common iliac vein

Testicular vein and artery

Superior pubic ramus

Bladder

SECTION THROUGH MALE BLADDER

SECTION THROUGH BOWMAN'S CAPSULE

Distal convoluted tubule

Efferent arteriole

Afferent arteriole

Basement membrane of Bowman's capsule

Bowman's space

Bowman's capsule

Glomerulus

Proximal convoluted tubule

Peritoneum

Right ureter

Urachus

Left ureter

Transitional cell mucosa

Right ureteric orifice

Internal urethral orifice

Muscle layer

Left ureteric orifice

Trigone

Prostate gland

Internal urethral sphincter muscle

Urethra

Reproductive system

SEX ORGANS LOCATED IN THE PELVIS create new human lives. Each month a ripe egg is released from one of the female's ovaries into a fallopian tube leading to the uterus (womb), a muscular pear-sized organ. A male produces minute tadpole-like sperm in two oval glands called testes. When the male is ready to release sperm into the female's vagina, many millions pass into his urethra and leave his body through the fleshy penis. The sperm travel up through the vagina into the uterus and one sperm may enter and fertilize an egg. The fertilized egg becomes embedded in the uterus wall and starts to grow into a new human being.

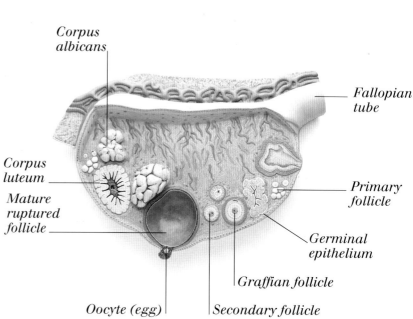

Corpus albicans

Fallopian tube

Corpus luteum

Mature ruptured follicle

Primary follicle

Germinal epithelium

Graffian follicle

Oocyte (egg)

Secondary follicle

SECTION THROUGH FEMALE PELVIC REGION

Ovary

Fundus of uterus

Uterus (womb)

Cervix (neck of uterus)

Os

Rectum

Vagina

Anus

Perineum

Introitus (vaginal opening)

Ureter

Ampulla of fallopian tube

Fimbria of fallopian tube

Isthmus of fallopian tube

Bladder

Pubic symphysis

Urethra

Clitoris

External urinary meatus

Labia minora

Labia majora

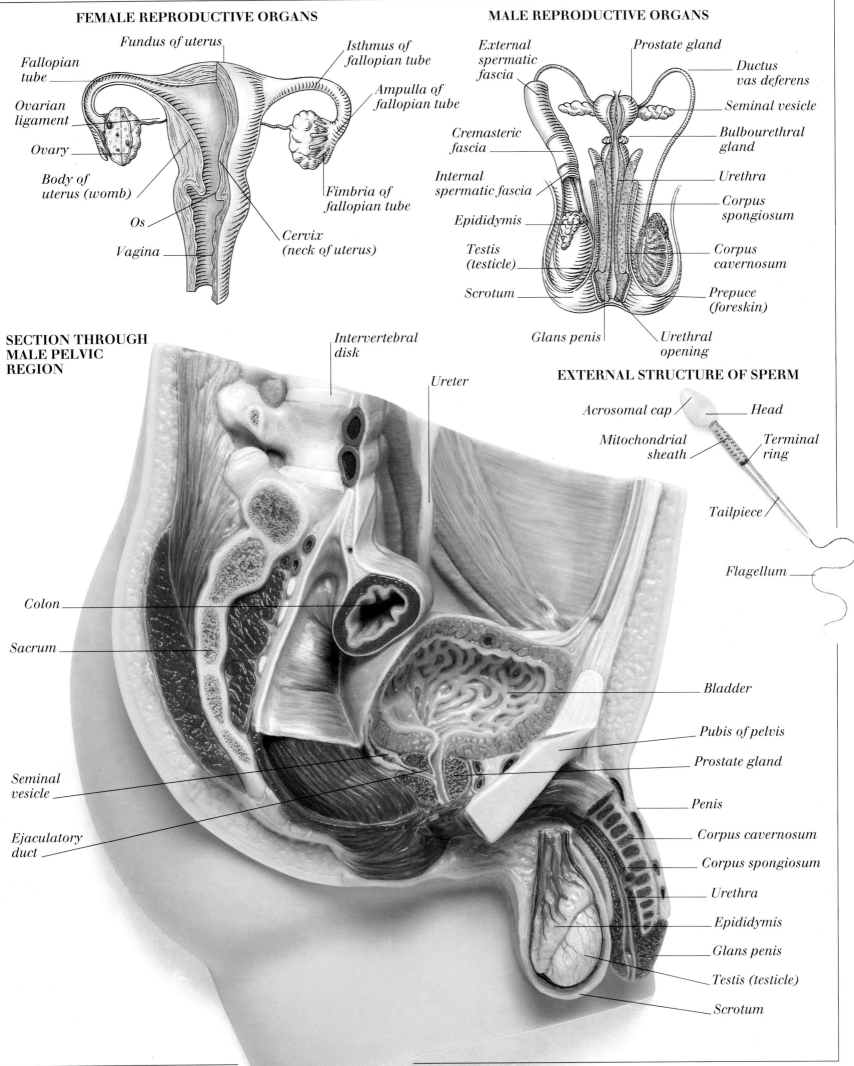

FEMALE REPRODUCTIVE ORGANS

Fallopian tube
Fundus of uterus
Isthmus of fallopian tube
Ovarian ligament
Ampulla of fallopian tube
Ovary
Body of uterus (womb)
Os
Fimbria of fallopian tube
Cervix (neck of uterus)
Vagina

MALE REPRODUCTIVE ORGANS

External spermatic fascia
Prostate gland
Ductus vas deferens
Cremasteric fascia
Seminal vesicle
Bulbourethral gland
Internal spermatic fascia
Urethra
Epididymis
Corpus spongiosum
Testis (testicle)
Corpus cavernosum
Scrotum
Prepuce (foreskin)
Glans penis
Urethral opening

SECTION THROUGH MALE PELVIC REGION

Intervertebral disk
Ureter
Colon
Sacrum
Seminal vesicle
Ejaculatory duct

EXTERNAL STRUCTURE OF SPERM

Acrosomal cap
Head
Mitochondrial sheath
Terminal ring
Tailpiece
Flagellum

Bladder
Pubis of pelvis
Prostate gland
Penis
Corpus cavernosum
Corpus spongiosum
Urethra
Epididymis
Glans penis
Testis (testicle)
Scrotum

57

Development of a baby

A FERTILIZED EGG IS NOURISHED AND PROTECTED as it develops into an embryo and then a fetus during the 40 weeks of pregnancy. The placenta, a mass of blood vessels implanted in the uterus lining, delivers nourishment and oxygen, and removes waste through the umbilical cord. Meanwhile, the fetus lies snugly in its amniotic sac, a bag of fluid that protects it against any sudden jolts. In the last weeks of the pregnancy, the rapidly growing fetus turns head down: a baby ready to be born.

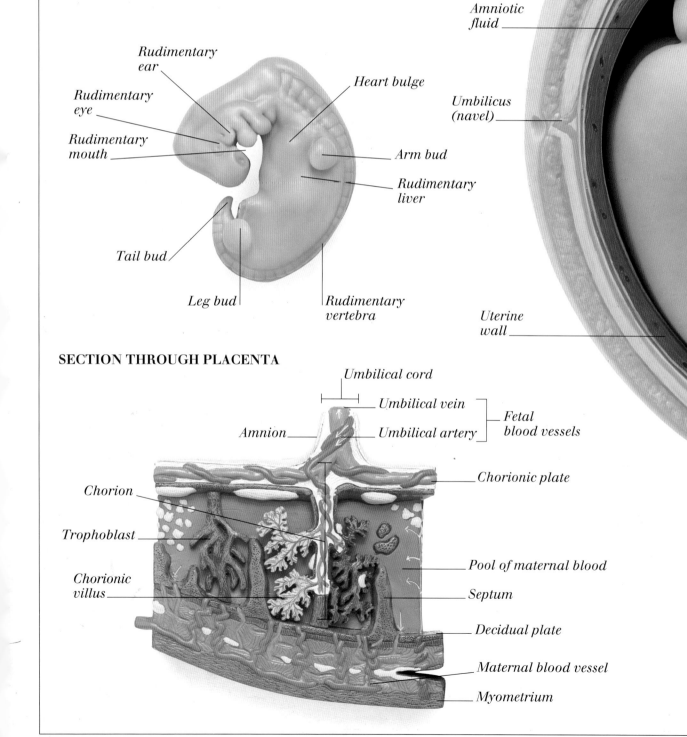

EMBRYO AT FIVE WEEKS

Rudimentary ear

Rudimentary eye

Rudimentary mouth

Heart bulge

Arm bud

Rudimentary liver

Tail bud

Leg bud

Rudimentary vertebra

Amniotic fluid

Umbilicus (navel)

Uterine wall

Fetus

SECTION THROUGH PLACENTA

Umbilical cord

Umbilical vein

Umbilical artery

Fetal blood vessels

Amnion

Chorion

Trophoblast

Chorionic villus

Chorionic plate

Pool of maternal blood

Septum

Decidual plate

Maternal blood vessel

Myometrium

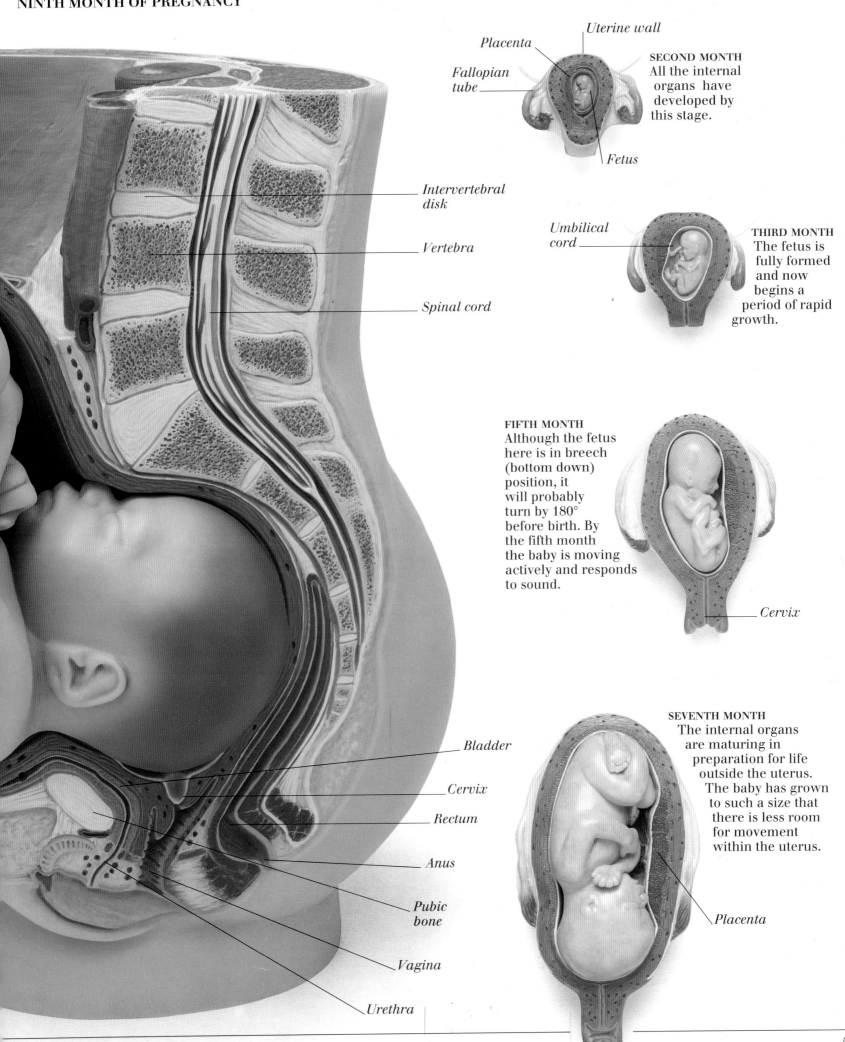

SECTION THROUGH PELVIS IN NINTH MONTH OF PREGNANCY

THE DEVELOPING FETUS

Uterine wall

Placenta

Fallopian tube

SECOND MONTH
All the internal organs have developed by this stage.

Fetus

Intervertebral disk

Vertebra

Spinal cord

Umbilical cord

THIRD MONTH
The fetus is fully formed and now begins a period of rapid growth.

FIFTH MONTH
Although the fetus here is in breech (bottom down) position, it will probably turn by 180° before birth. By the fifth month the baby is moving actively and responds to sound.

Cervix

Bladder

Cervix

Rectum

SEVENTH MONTH
The internal organs are maturing in preparation for life outside the uterus. The baby has grown to such a size that there is less room for movement within the uterus.

Anus

Pubic bone

Placenta

Vagina

Urethra

Index

Acknowledgments

Dorling Kindersley would like to thank :
Derek Edwards and Dr Martin Collins, British School of Osteopathy for skeletal material and advice; Dr M.C.E. Hutchinson, Department of Anatomy, United Medical and Dental Schools of Guy's and St Thomas' Hospitals for resin casts, additional skeletal material, and advice; models Barry O'Rorke (Bodyline Agency) and Pauline Swaine (MOT Model Agency).

Additional editorial assistance:
Susan Bosanko, Candace Burch, Deirdre Clark, Paul Docherty, Edwina Johnson, David Lambert, Gail Lawther, Dr Robert Youngson

Additional models
Bodyline, Donkin Models, Gordon Models, Morrison Frederick

Additional photography:
Dave Rudkin

Illustrators:
Simone End, Roy Flooks, David Gardner, Mick Gillah, Dave Hopkins, Linden Artists, John Woodcock

Index:
Dr Robert Youngson

Picture credits:
a=above, b=below, c=center, J=jacket, l=left, m=middle, r=right, t=top
Biophoto Associates: pages 13ca, cra, 24cbc, cbm, 26tr
KeyMed Ltd: 44bl, 45bl, bcl
Dr D.N. Landon (Institute of Neurology): 24bl, br
Life Science Images (Ron Boardman): 40bl, br
National Medical Slide Bank: 13cr

Science Photo Library: 10brc, 32; /Michael Abbey: 21t; /Agfa: Jct, 16tl; /Biophoto Associates: 13crb; /Dr Jeremy Burgess: 31bcl; /CNRI:10tl, cl, c, cr, bl, clb, crb, blc, br, 13cb, 31bcr, 34tl, 45bcr, 51tr, cra, 54tl; / Dr Brian Eyden: 24cbr; /Professor C. Ferlaud: 41clb; / Simon Fraser; 10 bcl; /Eric Grave: 13br; /Jan Hinsch: 21tc; /Manfred Kage: 13c, 31br, 33b; /Astrid and Hans-Freider Michler: 13tr; /NIBSC: 51br; /Omikron: 40bc; /David Scharf: 31bl; /Dr Klaus Schiller: 44bcl, bcr, br; /Secchi-Lecaque/Roussel-UCLAF/CNRI: 13tc, 51crb;/Stammers/Thompson: 26tl; /Sheila Terry: 30tl
Dr Christopher B. Williams (St Mark's Hospital): 45br
Dr Robert Youngson: 37cr
Zefa: 13bc; /H. Sochurek: Jcb, 6tl, 10cb, bcr, 48tl, 52tl

Picture research:
Sandra Schneider